Name ______________________ Class ______________ Date ____________

Skills Worksheet

Directed Reading

AF781456

Section: The Fish Body

Complete each statement by underlining the correct term or phrase in the brackets.

1. Fishes normally get oxygen from [the oxygen in water molecules / the oxygen dissolved in water].

2. Single-loop blood circulation means blood is pumped from the heart to the capillaries of the gills, through the body, and then to the [heart / capillaries of the body].

3. All fishes have an internal skeleton made of bone or [silicates / cartilage].

4. In fishes, fertilization generally takes place [externally / internally].

5. In a process called [spawning / fertilization], male and female gametes are released near each other in the water.

6. The eggs of sharks, skates, and rays are fertilized [inside / outside] the female's body.

Complete each statement by writing the correct term or phrase in the space provided.

7. Each gill is made of two rows of ______________________

______________________ .

8. At the rear end of the cheek cavity is an opening called a(n)

______________________ ______________________ .

9. In ______________________ ______________________ , water passing across the gills and the blood circulating in the capillary networks through the gills flow in opposite directions.

In the space provided, write the letter of the description that best matches the term or phrase.

______ **10.** sinus venosus

______ **11.** atrium

______ **12.** ventricle

______ **13.** conus arteriosus

a. a chamber that is a thick-walled pump with enough muscle to force blood to flow through the capillaries

b. a collection chamber that reduces the resistance of blood flow into the heart

c. a chamber that is a second pump; smooths pulsations and adds more force

d. a chamber that is filled with blood from the sinus venosus; has thin, muscular walls

Read each question, and write your answer in the space provided.

14. Why is maintaining water balance a different process for freshwater fishes than for marine fishes?

____

____

____

____

____

____

____

15. What are nephrons?

____

____

____

____

Name ______________________ Class ______________ Date ____________

Skills Worksheet

Directed Reading

Section: Today's Fishes

Complete each statement by writing the correct term or phrase in the space provided.

1. Most bony fishes have a hard plate called a(n) ______________________ that covers the gills on each side of the head.

2. The ______________________ ______________________ is a specialized sensory system that extends along each side of a bony fish's body.

3. By adjusting the gas content of the ______________________

______________________ , bony fishes can regulate their buoyancy.

Read each question, and write your answer in the space provided.

4. Describe the characteristics that are shared by lampreys and hagfishes.

____

____

____

____

____

____

5. Why are sharks such fast swimmers?

____

____

____

____

____

6. What are the distinguishing characteristics of rays and skates?

____

____

____

____

7. How do bony fishes benefit from having opercula?

____

____

____

____

8. How do bony fishes benefit from the lateral line?

____

____

____

____

____

9. Where does the gas in a bony fish's swim bladder come from?

____

____

10. What are teleosts?

____

____

____

____

____

11. What are lobe-finned bony fishes?

____

____

____

____

____

Name ______________________ Class ______________ Date ______________

Skills Worksheet

Directed Reading

Section: Amphibians

Read each question, and write your answer in the space provided.

1. What are the three orders of living amphibians?

____

____

____

____

2. What are five key characteristics of amphibians?

____

____

____

3. How do amphibians benefit from having pulmonary veins?

____

____

____

____

____

4. What is the role of the septum?

____

____

____

____

____

Complete each statement by writing the correct term or phrase in the space provided.

5. In amphibians, a pair of blood vessels called the ______________________ ______________________ carries oxygen-rich blood from the lungs to the heart.

6. In amphibians, a dividing wall called the ______________________ separates the atrium into right and left halves.

7. A(n) ______________________ is an internal, baglike respiratory organ that allows oxygen and carbon dioxide to be exchanged between the air and the bloodstream.

Complete each statement by underlining the correct term or phrase in the brackets.

8. Toads have [shorter / longer] legs than do frogs.

9. The skin of frogs is [smooth / covered with bumps].

10. The adult frog [hatches from an egg / is transformed from a tadpole].

11. Frogs reproduce by [external / internal] fertilization.

12. The female of a few species of salamanders achieves fertilization by picking up a sperm packet deposited somewhere by a male and placing the packet in her [cloaca / tympanic membrane].

13. Salamanders lay their eggs in water or in [moist places / dry sand].

14. Caecilians are legless, worm-shaped [jawless fishes / amphibians] that live on small invertebrates found in soil.

Name ______________________ Class ______________ Date ____________

Skills Worksheet

Active Reading

Section: The Fish Body

Read the passage below. Then answer the questions that follow.

The major respiratory organ of a fish is the gill. Each gill is made up of rows of **gill filaments,** which hang like curtains between a fish's mouth and cheeks. At the rear of the cheek cavity is an opening called a **gill slit.** As a fish swallows water, it forces the water from the mouth, over the gills, and out through the gill slits. This swallowing procedure is the core of a great change in gill design shown by fishes—countercurrent flow. In **countercurrent flow,** water passes over the gills in one direction as blood flows in the opposite direction through capillaries in the gills. Countercurrent flow ensures that oxygen diffuses into the blood over the entire length of the capillaries in the gills. Due to this arrangement, the gills of bony fishes can extract up to 85 percent of the oxygen in the water passing over them.

SKILL: READING EFFECTIVELY

Read each question, and write your answer in the space provided.

1. Why does a fish "swallow" water?

____

____

2. How are gills and gill filaments related?

____

____

3. How does countercurrent flow increase the efficiency of a bony fish's gills?

____

____

In the space provided, write the letter of the term or phrase that best completes the statement.

______ **4.** The major respiratory organ of a fish is the

a. lung.

b. gill filament.

c. gill.

d. sinus venosus.

Name ______________________ Class ______________ Date ____________

Skills Worksheet

Active Reading

Section: Today's Fishes

Read the passage below. Then answer the questions that follow.

In addition to their strong, internal skeleton made completely of bone, bony fishes have a series of unique structural adaptations that contribute greatly to their success.

Lateral line system: Bony fishes have a fully developed lateral line system. The **lateral line** is a specialized sensory system that extends along each side of a bony fish's body. As moving water presses against the lateral line, nerve impulses from ciliated sensory cells within it permit the fish to perceive its position and rate of movement. The lateral line system enables a fish to detect a motionless object by the movement of water deflected by that object.

Gill cover: Most bony fishes have a hard plate called an **operculum** that covers the gills on each side of the head. Movements of certain muscles and the opercula permit a bony fish to draw water over the gills, enabling the fish to take in oxygen. By using this mechanism, most bony fishes can move water over their gills while remaining stationary in the water. In other words, a bony fish doesn't have to swim forward with its mouth open to move water over its gills. This ability to respire without swimming enables a bony fish to conserve energy that can be spent chasing after prey and escaping from predators.

Swim bladder: Bony fishes contain a special gas sac called a **swim bladder.** By adjusting the gas content of the swim bladder, bony fishes can regulate their buoyancy. As the bladder fills, the fish rises, and as the bladder empties, the fish sinks. The swim bladder of early bony fishes was connected to the throat, and they could gulp air to fill it. The swim bladder of modern bony fishes does not have a direct passage to the mouth. Instead, gas is released from or absorbed into their bloodstream to fill or empty the bladder.

SKILL: READING EFFECTIVELY

Read each question, and write your answer in the space provided.

1. What three structural adaptations are found in bony fishes?

____

____

2. What is an operculum?

____

____

3. How does the ability to respire while stationary aid a bony fish?

____

____

____

4. What is the lateral line?

____

____

5. What two abilities does a bony fish have due to its lateral line system?

____

____

6. What is a swim bladder?

____

____

7. How did early bony fishes fill their swim bladder?

____

____

____

In the space provided, write the letter of the term or phrase that best completes the statement.

______ **8.** As the swim bladder of a bony fish empties, the fish

- **a.** rises.
- **b.** moves more rapidly.
- **c.** sinks.
- **d.** moves more slowly.

Name ______________________ Class ______________ Date ____________

Skills Worksheet

Active Reading

Section: Amphibians

Read the passage below. Then answer the questions that follow.

Most amphibians share five key characteristics.

Legs: The evolution of legs was an important adaptation for living on land. Frogs, toads, salamanders, and newts have four legs. Caecilians lost their legs during the evolutionary course of adapting to a burrowing existence.

Lungs: Although larval amphibians have gills, most adult amphibians breathe with a pair of lungs. Lungless salamanders are an exception.

Double-loop circulation: Two large veins called pulmonary veins return oxygen-rich blood from the lungs to the heart. The blood is then pumped to the body tissues at a much higher pressure than in the fish heart.

Partially divided heart: The atrium of the amphibian heart is divided into left and right sides, but the ventricle is not. A mixture of oxygen-rich and oxygen-poor blood is delivered to the amphibian's body tissues.

Cutaneous respiration: Most amphibians supplement their oxygen intake by respiring directly through their moist skin. Cutaneous respiration (skin breathing) limits the maximum body size of amphibians because it is efficient only when there is a high ratio of skin surface area to body volume.

SKILL: READING EFFECTIVELY

Read each question, and write your answer in the space provided.

1. Why do caecilians lack legs?

____

____

____

2. How is blood delivered to an amphibian's body tissues?

____

____

____

____

3. What is cutaneous respiration?

____

____

4. How does cutaneous respiration limit the body size of amphibians?

____

____

____

In the space provided, write the letter of the term or phrase that best completes the statement.

______ **5.** All of the following amphibians have four legs EXCEPT
- **a.** lungless salamanders.
- **b.** newts.
- **c.** toads.
- **d.** caecilians.

Name ______________________ Class ______________ Date ______________

Skills Worksheet

Vocabulary Review

Complete the crossword puzzle using the clues provided.

ACROSS

2. In a fish's gills, oxygen diffuses into the blood over the entire length of the capillaries due to __________ flow.

4. covers the gills on each side of the head

7. The __________ line extends along each side of a bony fish's body and allows the fish to perceive its position and rate of movement.

8. an internal, baglike respiratory organ

11. A swim __________ enables bony fishes to regulate their buoyancy.

DOWN

1. Oxygen-rich blood is carried from an amphibian's lungs to its heart by the __________ veins.

3. tubelike unit in a kidney

5. a fish with a completely symmetrical tail, highly mobile fins, and very thin scales

6. Each gill is made up of rows of fingerlike projections called gill __________ .

9. A __________ slit is an opening at the rear of the cheek cavity.

10. separates the amphibian atrium into right and left halves

Skills Worksheet

Science Skills

Interpreting Diagrams

The diagram below shows the structure of a fish heart. The arrows indicate the direction of blood flow into and out of the heart. Use the diagram to answer questions 1–6.

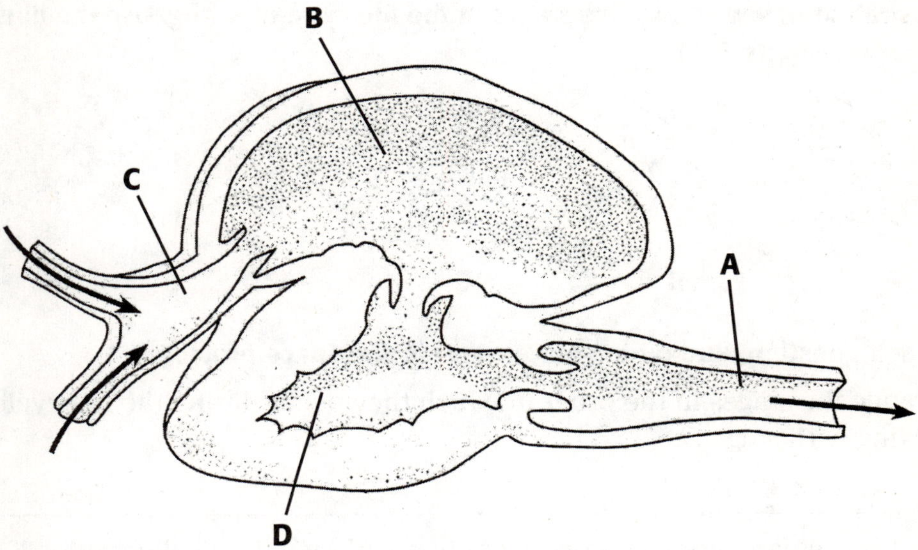

Read each question, and write your answer in the space provided.

1. Identify chambers *A*, *B*, *C*, and *D* in the diagram.

____

____

2. What is the function of chamber *A*?

____

____

3. What is the function of chamber *C*?

____

____

4. Where does blood go when it leaves chamber *A*?

____

5. Where does blood that enters chamber *C* come from?

____

6. How do the walls of chamber *D* enable this chamber to carry out its function?

The illustration below shows five stages in the life cycle of a frog. Use the illustration to answer questions 7–11.

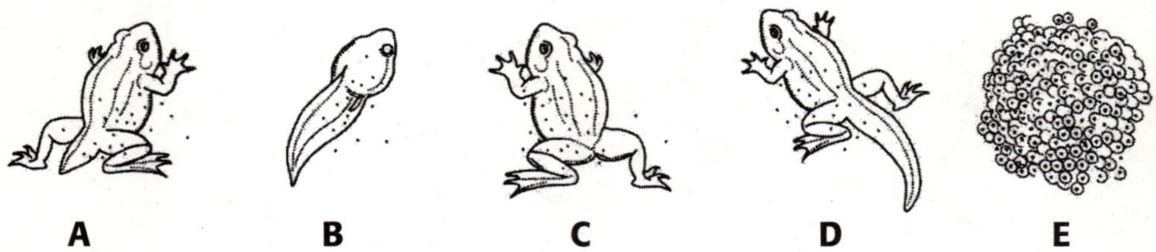

Read each question, and write your answer in the space provided.

7. Arrange the stages in the order in which they occur during the life cycle, starting with stage *E*.

8. What is the name for the process of change shown in this illustration?

9. What is stage *B* called?

10. List four differences between stage *B* and an adult frog.

11. Describe the process by which stage *E* is produced.

Name ______________________________ Class ______________ Date ____________

Skills Worksheet

Concept Mapping

Using the terms and phrases provided below, complete the concept map showing the classification of fishes.

cartilaginous	jawless	sharks
coelacanth	lungfishes	skates
hagfishes	ray-finned fishes	

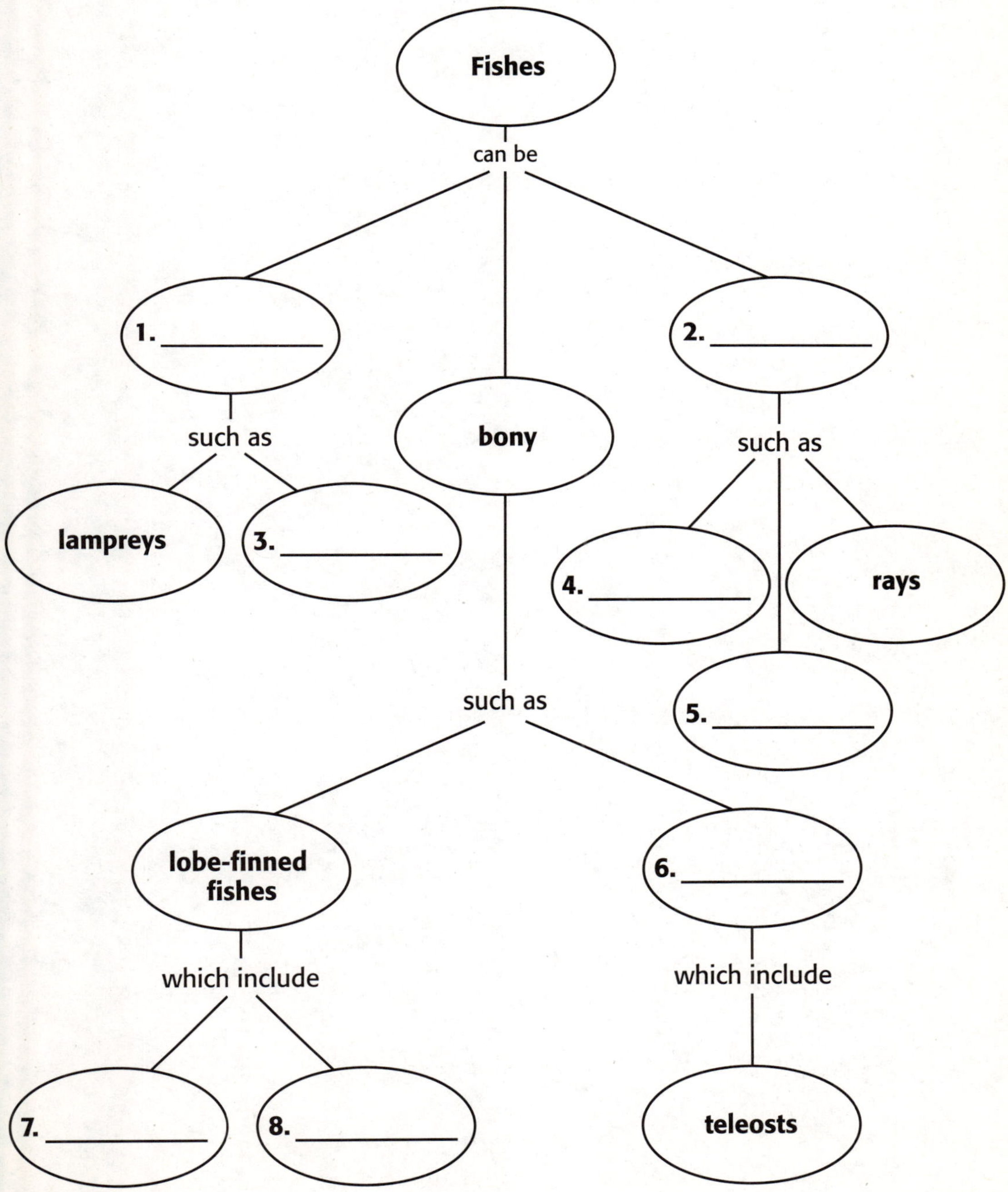

Name ______________________ Class ______________ Date ____________

Skills Worksheet

Critical Thinking

Work-Alikes

In the space provided, write the letter of the term or phrase that best describes how each numbered item functions.

______ **1.** hagfishes

______ **2.** swim bladder

______ **3.** lobe-finned structure

______ **4.** cutaneous respiration

a. buoyancy tank in a submarine

b. "vultures of the sea"

c. future legs

d. "skin breathing"

Cause and Effect

In the space provided, write the letter of the term or phrase that best matches each cause or effect given below.

Cause	Effect
5. countercurrent flow	______________
6. ______________	strong but lightweight skeleton of sharks
7. lateral line system	______________
8. frog egg external fertilization	______________

a. cartilage is strengthened by calcium carbonate

b. 85 percent extraction of oxygen by fish gills

c. tadpoles

d. position and rate of movement perceived

Trade-offs

In the space provided, write the letter of the bad news item that best matches each numbered good news item below.

Good News

______ **9.** Sharks continually lose and replace teeth.

______ **10.** Bony joints are found in lobed-finned fishes.

______ **11.** Fish have 20 times as much oxygen available to them in air compared to water.

______ **12.** Lungs evolved in amphibians.

______ **13.** Some salamanders manage to live in dry areas.

Bad News

a. They may lose 20,000 in their lifetime.

b. Gills cannot function out of water.

c. They contain a mixture of oxygen-rich and oxygen-poor air.

d. Rays are found at the tips.

e. They must remain inactive during the day.

Linkages

In the spaces provided, write the letters of the two terms or phrases that are linked together by the term or phrase in the middle. The choices can be placed in any order.

14. ______ blood circulating in gills from opposite directions ______

15. ______ atrium and ventricle ______

16. ______ water pumps over fish gills ______

17. ______ tadpoles hatch ______

18. ______ salamanders and newts ______

a. sinus venosus

b. ability to respire without swim ming

c. external fertilization of frog eg

d. operculum movement

e. frogs and toads

f. oxygen diffuses into blood ove entire length of the gills

g. caecilians

h. metamorphosis

i. moving water passing across fish gills

j. conus arteriosus

Analogies

An analogy is a relationship between two pairs of terms or phrases written as a : b :: c : d. The symbol : is read as "is to," and the symbol :: is read as "as." In the space provided, write the letter of the pair of terms or phrases that best completes the analogy shown.

______**19.** gills : filaments ::

a. kidneys : nephrons

b. kidneys : chambers

c. kidneys : filaments

d. hearts : filaments

______**20.** marine fishes : small amounts of urine ::

a. freshwater fishes : little urine

b. freshwater fishes : small amounts of dilute urine

c. freshwater fishes : small amounts of concentrated urine

d. freshwater fishes : large amounts of dilute urine

______**21.** bony fishes : movements of opercula ::

a. sharks : using swim bladder

b. amphibians : lungs and skin

c. sharks : movement of opercula

d. bony fishes : rows of teeth

______**22.** salamanders : tails ::

a. anurans : tails

b. caecilians : legs

c. caecilians : no legs

d. frogs : no legs

Name ______________________ Class ______________ Date ______________

Skills Worksheet

Test Prep Pretest

In the space provided, write the letter of the term or phrase that best completes each statement or best answers each question.

______ **1.** The major respiratory organ of a fish is the
a. swim bladder. **c.** gill.
b. lung. **d.** nephron.

______ **2.** Depending on the species, fish can reproduce through
a. internal fertilization. **c.** conjugation.
b. spawning. **d.** Both (a) and (b)

______ **3.** Lampreys and hagfishes are the only remaining
a. jawless fishes. **c.** lobe-finned fishes.
b. cartilaginous fishes. **d.** bony fishes

______ **4.** Members of the order Apoda do NOT
a. use cutaneous respiration. **c.** have legs.
b. lay eggs. **d.** bear live young.

______ **5.** Compared with that of a fish, a frog's cerebrum is
a. less complex. **c.** smaller.
b. more complex. **d.** Both (b) and (c)

Questions 6 and 7 refer to the figure at right, which shows the structure of a bony fish.

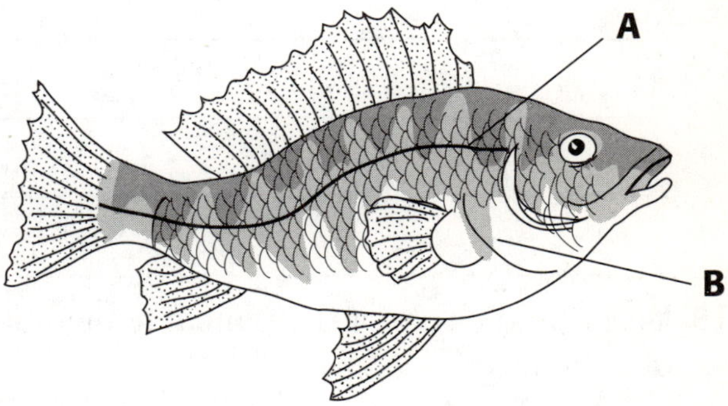

______ **6.** The structure labeled *A* is the
a. dorsal fin.
b. gill filament.
c. operculum.
d. lateral line.

______ **7.** The structure labeled *B* is called the
a. pectoral fin. **c.** pelvic fin.
b. operculum. **d.** jaw.

Complete each statement by writing the correct term or phrase in the space provided.

8. Fishes respire by using ______________________.

9. A shark's teeth are actually modified ______________________.

10. In an amphibian, the ____________ ____________ carry oxygen-rich blood from the lungs to the heart.

11. All fishes have a(n) ____________ ____________ that surrounds the spinal cord.

12. In amphibians, one circulatory loop carries blood from the heart to the

____________ , while a second loop carries blood to the rest of the body.

13. The large size of the ____________ ____________ in the yellow perch's brain indicates the importance of vision to this fish.

14. Tubelike units called ____________ in the kidneys of a fish regulate salt and water balance.

15. Salamanders are members of the order ____________ .

16. The pattern of movement of water and blood through a fish's gills is called

____________ ____________ .

Read each question, and write your answer in the space provided.

17. What is countercurrent flow, and why is it important to a fish?

____

____

____

____

____

18. Describe the role of the tympanic membrane in a leopard frog's sense of balance.

____

____

____

____

____

19. How do lampreys and hagfishes feed?

____

____

____

20. What information does a bony fish get from its lateral line system?

____

____

____

Name ______________________ Class ______________ Date ____________

Assessment

Quiz

Section: The Fish Body

In the space provided, write the letter of the description that best matches the term or phrase.

______ **1.** clasper

______ **2.** gill filament

______ **3.** gill slit

______ **4.** nephron

______ **5.** sinus venosus

______ **6.** conus arteriosus

a. opening at the rear of a fish's cheek cavity

b. organ used by a male shark to insert sperm into a female

c. heart chamber that smooths pulsations of the blood

d. collection chamber of a fish's heart

e. structure through which gases enter and leave a fish's blood

f. tubelike unit in a kidney

In the space provided, write the letter of the term or phrase that best completes each statement or best answers each question.

______ **7.** Because the bodies of freshwater fishes contain more ions than the surrounding water, they
- **a.** excrete concentrated urine.
- **b.** need to take in salts from the environment.
- **c.** tend to take in water through osmosis.
- **d.** Both (b) and (c)

______ **8.** Which of the following is NOT a characteristic of most fishes?
- **a.** gills
- **b.** vertebral column
- **c.** single-loop circulation
- **d.** external skeleton

______ **9.** Fish "swallow" water to
- **a.** force it from the mouth and over the gills.
- **b.** steady themselves in the water.
- **c.** feed.
- **d.** All of the above

______ **10.** In most fishes, blood circulates from the heart to the
- **a.** rest of the body, to the gills, and back to the heart.
- **b.** gills, to the rest of the body, and back to the heart.
- **c.** gills, back to the heart, and to the rest of the body.
- **d.** lungs, to the rest of the body, and back to the heart.

Name ______________________ Class ______________ Date ______________

Assessment

Quiz

Section: Today's Fishes

In the space provided, write the letter of the description that best matches the term or phrase.

______ **1.** hagfish

______ **2.** skate

______ **3.** lateral line

______ **4.** coelacanth

______ **5.** anal fin

______ **6.** lamprey

a. structure that keeps a fish from veering sideways as it swims

b. cartilaginous fish with a flattened body

c. parasitic fish with a structure like a suction cup around its mouth

d. jawless fish that produces slime when threatened

e. lobe-finned bony fish

f. sensory system on each side of a bony fish's body

In the space provided, write the letter of the term or phrase that best completes each statement or best answers each question.

______ **7.** To rise in the water, a bony fish fill its swim bladder with gas from
- **a.** its gills.
- **b.** its bloodstream.
- **c.** the surrounding water.
- **d.** its lungs.

______ **8.** Jawless fishes have which of the following?
- **a.** unpaired fins
- **b.** scales
- **c.** bony skeletons
- **d.** swim bladders

______ **9.** A bony fish uses its opercula to
- **a.** regulate its salt and water balance.
- **b.** detect vibrations in the water.
- **c.** draw water over its gills.
- **d.** coordinate muscle activity.

______ **10.** The yellow perch is an example of a
- **a.** lobe-finned fish.
- **b.** jawless fish.
- **c.** cartilaginous fish.
- **d.** teleost.

Name ______________________ Class ______________ Date ______________

Assessment

Quiz

Section: Amphibians

In the space provided, write the letter of the description that best matches the term or phrase.

_____ **1.** cutaneous respiration

_____ **2.** metamorphosis

_____ **3.** caecilian

_____ **4.** tadpole

_____ **5.** lung

_____ **6.** salamander

a. internal, baglike respiratory organ

b. process of dramatic physical change

c. long-tailed amphibian with an elongated body

d. skin breathing

e. legless, burrowing amphibian

f. fishlike, larval form of a frog

In the space provided, write the letter of the term or phrase that best completes each statement or best answers each question.

_____ **7.** In a modern amphibian's heart,
a. neither the atrium nor the ventricle is divided.
b. both the atrium and the ventricle are divided.
c. only the ventricle is divided.
d. only the atrium is divided.

_____ **8.** Frogs and toads are members of which order?
a. Anura
b. Apoda
c. Urodela
d. Cloaca

_____ **9.** Because of their tympanic membrane, leopard frogs hear
a. well in both water and air.
b. well in water but not in air.
c. well in air but not in water.
d. poorly in both water and air.

_____ **10.** A frog's pulmonary veins carry blood from the
a. body organs.
b. atria.
c. lungs.
d. ventricle.

Name ______________________ Class ______________ Date ____________

Assessment

Chapter Test

Fishes and Amphibians

In the space provided, write the letter of the description that best matches the term or phrase.

______ 1. spawning

______ 2. pulmonary vein

______ 3. cloaca

______ 4. countercurrent flow

______ 5. teleost

______ 6. operculum

______ 7. septum

a. wall that divides the atrium into right and left halves

b. release of male and female gametes near each other in the water

c. opposite movement of water and blood through a fish's gills

d. structure that carries blood from the lungs to the heart

e. hard plate that covers the gills on each side of the head

f. most advanced of the ray-finned bony fishes

g. chamber through which urine, gametes, and undigested material pass

In the space provided, write the letter of the term or phrase that best completes each statement or best answers each question.

______ **8.** Gills, single-loop circulation, and a vertebral column are found in
a. bony fishes.
b. cartilaginous fishes.
c. jawless fishes.
d. All of the above

______ **9.** Members of the category Osteichthyes
a. have skeletons made of bone.
b. do not have jaws.
c. include rays and skates.
d. All of the above

______ **10.** The living jawless fishes are
a. lampreys and sharks.
b. sharks and rays.
c. hagfishes and coelacanths.
d. lampreys and hagfishes.

______ **11.** Cartilaginous fishes have all of the following EXCEPT
a. kidneys.
b. internal fertilization.
c. a swim bladder.
d. gill slits.

______ **12.** Freshwater fishes tend to take in water through osmosis and therefore excrete which of the following?
a. small amounts of concentrated urine
b. large amounts of concentrated urine
c. small amounts of dilute urine
d. large amounts of dilute urine

______**13.** Members of the order Anura
a. retain gills as adults.
b. have neither jaws nor legs.
c. undergo metamorphosis.
d. have small, bony scales embedded in their skin.

______**14.** A fish absorbs oxygen through its
a. gill slits.
b. swim bladder.
c. gill filaments.
d. mouth.

______**15.** A yellow perch propels itself forward with its
a. caudal fin.
b. pelvic fins.
c. pectoral fins.
d. dorsal fins.

______**16.** In a yellow perch, undigested food exits the body through the
a. nephron.
b. anus.
c. intestine.
d. ventral anal fin.

______**17.** The fish heart is an evolutionary advance for vertebrates because it
a. pumps blood through a double loop.
b. has a divided ventricle.
c. is stronger than the tubular hearts of early chordates.
d. has a divided atrium.

______**18.** Bony fishes regulate their buoyancy with their
a. pectoral fins.
b. swim bladder.
c. gills.
d. lateral line.

Questions 19 and 20 refer to the figure at right, which shows the lateral line system.

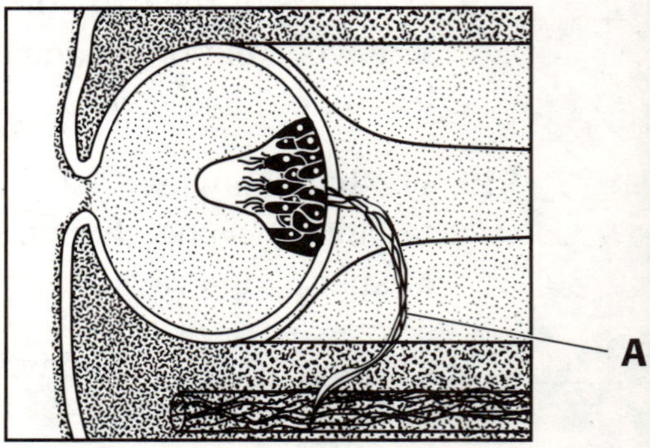

______**19.** The lateral line system enables a fish to
a. detect vibrations in the water.
b. see objects at a distance.
c. smell.
d. swim rapidly.

______**20.** The structure labeled *A* is a
a. cilium.
b. scale.
c. supporting cell.
d. nerve.

Name ______________________ Class ______________ Date ____________

Assessment

Chapter Test

Fishes and Amphibians

In the space provided, write the letter of the description that best matches the term or phrase.

______ **1.** tympanic membrane

______ **2.** countercurrent flow

______ **3.** single-loop circulation

______ **4.** septum

______ **5.** spawning

______ **6.** urostyle

a. flow of blood from the heart to the gills to the rest of the body to the heart

b. method of accomplishing external fertilization in most fishes

c. structure that prevents complete mixing of blood

d. mechanism that makes respiration efficient in fish gills

e. eardrum of a leopard frog

f. structure made of fused vertebrae in a leopard frog

In the space provided, write the letter of the term or phrase that best completes each statement or best answers each question.

______ **7.** Which of the following is NOT a characteristic of teleosts?
- **a.** highly mobile fins
- **b.** very thin scales
- **c.** completely symmetrical tail
- **d.** skeleton made of cartilage

______ **8.** A bony fish uses its swim bladder to
- **a.** detect motionless objects.
- **b.** regulate its buoyancy.
- **c.** store urine produced by the kidneys.
- **d.** store bile produced by the liver.

______ **9.** Most salamanders have
- **a.** gills as adults.
- **b.** rough, dry skin.
- **c.** unpaired fins.
- **d.** an elongated body.

______ **10.** Marine fishes maintain salt and water balance by
- **a.** excreting large amounts of dilute urine.
- **b.** pumping excess ions into their body.
- **c.** drinking sea water.
- **d.** All of the above

_____ **11.** In a fish heart, the thick-walled chamber that contracts strongly is the
- **a.** sinus venosus.
- **b.** atrium.
- **c.** ventricle.
- **d.** conus arteriosus.

_____ **12.** All modern fishes have a vertebral column that
- **a.** surrounds the spinal cord.
- **b.** is made of cartilage.
- **c.** is made of bone.
- **d.** is part of an external skeleton.

Complete each statement by writing the correct term or phrase in the space provided.

13. By moving certain ______________________ and the ______________________, bony fishes can move water over their gills while remaining stationary.

14. Besides breathing with their lungs, amphibians also engage in ______________________ respiration.

15. The only jawless fishes that survive today are ______________________ and ______________________.

16. Urine, undigested food, and gametes all exit the body of a leopard frog through the ______________________ opening.

17. The ______________________ of a yellow perch grow throughout its life and can be used to estimate the ______________________ of the fish.

18. In amphibians, the ______________________ veins carry blood from the lungs to the heart.

19. The changes that transform a tadpole into an adult frog are called ______________________.

20. In the kidneys of a fish, tubelike units called ______________________ remove metabolic wastes from the blood.

Read each question, and write your answer in the space provided.

21. Which group of amphibians does not have legs? Explain why.

____

____

____

22. What enables a leopard frog to keep its skin moist? Why is moist skin important for this animal?

____

____

____

23. Explain the importance of the operculum to bony fishes.

____

____

____

24. Explain the roles of the caudal, dorsal, and pectoral fins in the movement of a yellow perch.

____

____

____

____

____

25. Trace the circulation of blood through the amphibian heart.

____

____

____

____

____

____

____

Name ______________________ Class ______________ Date ____________

Data Lab

DATASHEET FOR IN-TEXT LAB

Analyzing Ion Excretion in Fish

Background

A few species of fish, such as adult salmon, are able to move between salt-water and freshwater environments. The graph below shows the excreted ion concentration of a fish as it travels from one body of water to another. Examine the graph, and answer the analysis questions.

Ion Excretion in Fish

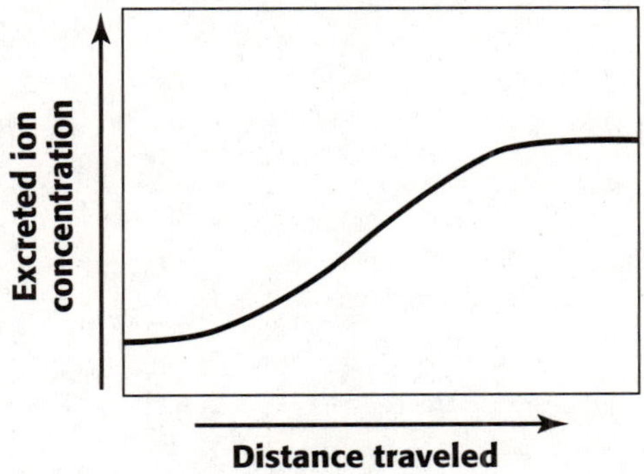

Analysis

1. **Determine** if the fish is losing or gaining ions by excretion as it travels.

____

2. **Critical Thinking**
   Inferring Conclusions Is the fish traveling from fresh water to salt water or from salt water to fresh water?

____

3. **Summarize** the reasoning you used to answer item 2.

____

____

____

____

____

____

Name ______________________ Class ______________ Date ____________

Quick Lab

DATASHEET FOR IN-TEXT LAB

Modeling the Action of a Swim Bladder

Most fish use a swim bladder to regulate their depth in water. As gas enters the swim bladder, the fish rises in the water. As gas is expelled, the fish sinks to a lower depth.

MATERIALS

- 100 mL beaker or small glass
- cold, clear, carbonated soft drink
- 2 very dry raisins

Procedure

1. Fill a 100 mL beaker with a cold, carbonated soft drink.
2. Drop two raisins into the beaker, and observe what happens over the next 5 minutes.

Analysis

1. **Describe** what happened after you dropped the raisins into the soft drink.

____

____

____

2. **Forming Hypotheses** Develop a hypothesis to explain your observations.

____

____

____

____

3. **Critical Thinking**
Analyzing Results How does the lifting of the raisins differ from the use of a swim bladder to control buoyancy?

____

____

____

____

4. Critical Thinking

Forming Reasoned Opinions Think about the energy you would have to expend to keep yourself in one position underwater. What advantage might a swim bladder provide to a fish?

____

____

____

____

____

Name ______________________ Class ______________ Date ____________

Exploration Lab

DATASHEET FOR IN-TEXT LAB

Observing a Live Frog

SKILLS

- Observing
- Relating

OBJECTIVES

- **Examine** the external features of a frog.
- **Observe** the behavior of a frog.
- **Explain** how a frog is adapted to life on land and in water.

MATERIALS

- live frog in a terrarium
- live insects (crickets or mealworms)
- 600 mL beaker
- aquarium half-filled with dechlorinated water

Before You Begin

Frogs, which are **amphibians,** are adapted for living on land and in water. For example, a frog's eyes have an extra eyelid called the **nictitating membrane.** This eyelid protects the eye when the frog is underwater and keeps the eye moist when the frog is on land. The smooth skin of a frog acts as a respiratory organ by exchanging oxygen and carbon dioxide with the air or water. The limbs of a frog enable it to move both on land and in water. In this lab, you will examine a live frog in both a terrestrial environment and an aquatic environment.

1. Write a definition for each boldface term in the paragraph above and for the following term: tympanic membrane. Use a separate sheet of paper.
2. You will be using the data table provided to record your data.
3. Based on the objectives for this lab, write a question you would like to explore about frogs.

____

____

Procedure

PART A: OBSERVING A FROG

1. Observe a live frog in a terrarium. Closely examine the external features of the frog. Make a drawing of the frog. Label the eyes, nostrils, tympanic membranes, front legs, and hind legs.

2. Watch the frog's movements as it breathes air into and out of its lungs. Record your observations.

Data Table	
Behavior/structure	**Observations**
Breathing	
Eyes	
Legs	
Response to food	
Response to noise	
Skin	
Swimming behavior	

3. Look closely at the frog's eyes, and note their location. Examine the upper and lower eyelids as well as a third transparent eyelid called a *nictitating membrane*. Describe how the eyelids move.

4. Study the frog's legs, and note the difference between the front and hind legs.

5. Place a live insect, such as a cricket or a mealworm, into the terrarium. Observe how the frog reacts.

6. Tap the side of the terrarium farthest from the frog, and observe the frog's response.

7. Place a 600 mL beaker in the terrarium. **CAUTION: Handle live frogs gently. Frogs are slippery! Do not allow a frog to injure itself by jumping from a lab table to the floor.** Carefully pick up the frog, and examine its skin. How does it feel? Now place the frog in the beaker. Cover the beaker with your hand, and carry it to a freshwater aquarium. Tilt the beaker, and gently lower it into the water until the frog swims out.

8. Watch the frog float and swim. Notice how the frog uses its legs to swim. Also notice the position of the frog's head. As the frog swims, bend down to view the underside of the frog. Then look down on the frog from above. Compare the color on the dorsal and ventral sides of the frog.

PART B: CLEANUP AND DISPOSAL

9. Dispose of broken glass in the designated waste containers. Put live animals in the designated containers. Do not pour chemicals down the drain or put lab materials in the trash unless your teacher tells you to do so.

10. Clean up your work area and all lab equipment. Return lab equipment to its proper place. Wash your hands thoroughly before you leave the lab and after you finish all work.

Analyze and Conclude

1. **Summarizing Information** How does a frog use its hind legs for moving on land and in water?

___

2. **Recognizing Relationships** How does the position of a frog's eyes benefit the frog while it is swimming?

___

___

3. **Analyzing Data** What features of an adult frog provide evidence that it has an aquatic life and a terrestrial life?

___

___

4. **Analyzing Methods** Were you able to determine in this lab how a frog hears? Explain.

___

___

___

5. **Inferring Conclusions** What can you infer about a frog's field of vision from the position of its eyes?

___

6. **Forming Hypotheses** How is the coloration on the dorsal and ventral sides of a frog an adaptive advantage?

___

___

___

___

7. **Further Inquiry** Write a new question about frogs that could be explored with another investigation.

___

___

Name ______________________ Class ______________ Date ____________

Exploration Lab

FIELD ACTIVITY

Observing Structure and Behavior in Fishes

Most fishes share several characteristics, including gills, fins, and scales, that distinguish them from other vertebrates. Beyond these shared characteristics, the 25,000 species of fishes that inhabit Earth's oceans, lakes, ponds, and rivers exhibit a great diversity of body styles and behaviors.

In this lab, you will observe the structural and behavioral adaptations of several species of fish at an aquarium. You will then consider how these adaptations help each species survive in its habitat.

OBJECTIVES

Observe body shapes, camouflage, feeding behavior, and schooling behavior in various species of fish.

Relate observed adaptations to the survival of each species.

MATERIALS

- clipboard and paper or notebook
- pencil or pen

Procedure

BODY SHAPES

1. Find a tank that contains trout, herring, barracuda, or tuna. If the aquarium does not have any of these fishes, ask your teacher for some alternatives.
2. Identify the species of fish in the tank. In **Table 1,** record the common and scientific names of the species.
3. Describe the shape of the fish's body. In **Table 1,** record your description.
4. Observe the activity of the fish. How does it spend most of its time? In **Table 1,** record your observations.
5. Repeat steps 2–4 with another species of fish from each of the following two groups: (1) flounder, sole, ray, or skate; and (2) eel or lamprey.

TABLE 1 BODY SHAPES AND ACTIVITY

Species	Body shape	Activity

CAMOUFLAGE

6. Find three examples of fish that use body colors, markings, or postures to blend with their surroundings.

7. For each species, record its common and scientific names in **Table 2,** and describe the type of camouflage it uses.

TABLE 2 CAMOUFLAGE

Species	Type of camouflage

FEEDING BEHAVIOR

8. Find three examples of fish that use different methods for feeding.

9. For each species, record its common and scientific names in **Table 3,** and describe its feeding method.

TABLE 3 FEEDING BEHAVIOR

Species	Type of feeding behavior

SCHOOLING BEHAVIOR

10. Look for a tank that contains a large group of fish. Decide whether or not the group is a school. Among the kinds of fish that form schools are anchovy, barracuda, bluefish, bonito, dace, herring, mackerel, and shark.

What criteria would help you decide whether a group of fish is a school?

____

____

11. Identify the species of fish in the school. In **Table 4,** record the common and scientific names of the species.

TABLE 4 SCHOOLING BEHAVIOR

Species	Number of individuals	Observations

12. Count or estimate the number of individuals in the school. Record the number in **Table 4.**

13. Observe the school carefully. Do the fish spend all of their time in the school, or does the school break up occasionally and then reform? If it breaks up and reforms, what appears to trigger school formation? Record your observations in **Table 4.**

14. Repeat steps 11–13 with another school of fish in a different tank.

Analysis

1. Identifying Relationships How are the body shapes of the fishes you observed related to the types of activities that those fishes engage in?

____

____

____

____

2. Recognizing Patterns What structural adaptations are associated with the different feeding behaviors that you observed?

____

____

____

____

Conclusions

1. **Drawing Conclusions** How does camouflage help fish to survive in the cases where you observed it?

2. **Applying Conclusions** Would you expect camouflage to be more important for fish that live in shallow water or for those that live in very deep water? Explain your reasoning.

3. **Making Predictions** Why might it be advantageous for fish to swim in schools?

Extensions

1. **Designing Experiments** Think of a question you would like to answer about fish structure or behavior. Design an experiment that could answer your question, using fish sold at a local pet store and materials available in your classroom.
2. **Research and Communications** Research another class of vertebrates in which individuals move in groups. Compare their movements with those of the fish you observed. Report the results of your research in an oral presentation.

Name ______________________ Class ______________ Date ______________

Skills Practice Lab

OBSERVATION

Frog Dissection

Amphibians, such as frogs, are examples of animals that have structural adaptations for living in or near water and for living on land. As tadpoles, they have gills, but adult frogs have lungs. Tadpoles have tails, but adult frogs have legs. Many changes occur in the metamorphosis of a tadpole. The structures that frogs have depend on the stage of the frog's life cycle. In this lab, you will dissect a frog in order to observe the external and internal structures of frog anatomy.

MATERIALS

- dissecting pins (6–10)
- dissecting tray
- dull probe or dissecting needle
- felt-tip marker
- forceps
- gloves
- lab apron
- paper towels
- plastic storage bag and twist tie
- preserved frog
- safety goggles
- scissors

OBJECTIVES

Describe the appearance of various organs found in a frog.

Name the organs that make up various systems of a frog.

Procedure

1. Put on safety goggles, gloves, and a lab apron.
2. Rinse and place a frog on a dissection tray with its dorsal side down. To determine the frog's sex, look at the hand digits, or fingers, on its forelegs. A male frog usually has thick pads on its "thumbs," which is one external difference between the sexes, as shown in **Figure 1.** Male frogs are also usually smaller than female frogs. Observe several frogs to see the difference between males and females.

FIGURE 1 HAND DIGITS OF MALE AND FEMALE FROG

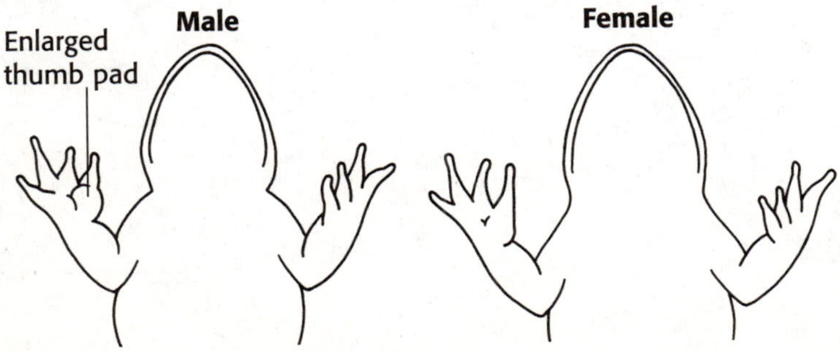

3. Review the following information about the frog's head.

On the outside of the frog's head are two *external nares*, or nostrils; two *tympani*, or eardrums; and two eyes, each of which has three lids. The third lid, called the *nictitating membrane*, is transparent.

4. Use **Figure 2** to locate and identify the external features of the head. Find the mouth, external nares, tympani, eyes, and nictitating membranes.

5. Review the following information about the frog's mouth.

Inside the mouth are two *internal nares*, or openings into the nostrils; two *vomerine teeth* in the middle of the roof of the mouth; and two *maxillary teeth* at the sides of the mouth. Also inside the mouth behind the tongue is the *pharynx*, or throat. In the pharynx, there are several openings: one into the *esophagus*, the tube into which food is swallowed; one into the *glottis*, through which air enters the *larynx*, or voice box; and two into the *Eustachian tubes*, which connect the pharynx to the ear.

FIGURE 2 EXTERNAL FEATURES OF FROG HEAD

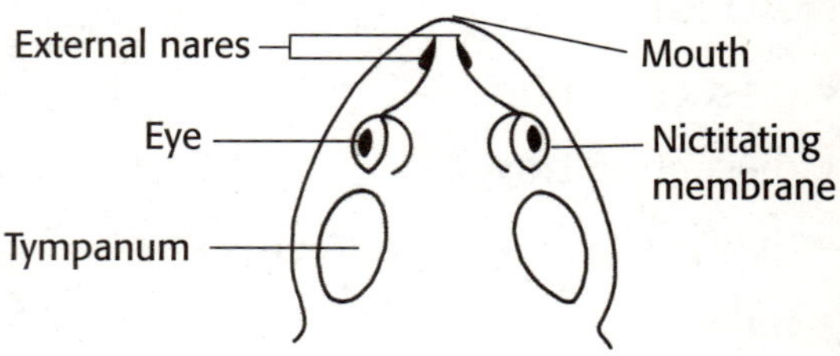

6. Turn the frog on its back, and pin down the legs. Cut the hinges of the mouth, and open it wide. Use **Figure 3** to locate and identify the structures inside the mouth. Use a dull probe to help find each part: the vomerine teeth, the maxillary teeth, the internal nares, the tongue, the openings to the Eustachian tubes, the esophagus, the pharynx, and the slitlike glottis.

7. Look for the opening to the frog's cloaca, located between the hind legs. Use forceps to lift the skin, and use scissors to cut along the center of the body from the cloaca to the lip. Turn back the skin, cut toward the side at each leg, and pin the skin flat. **Figure 4** shows how to make these cuts.

FIGURE 3 STRUCTURES OF FROG MOUTH

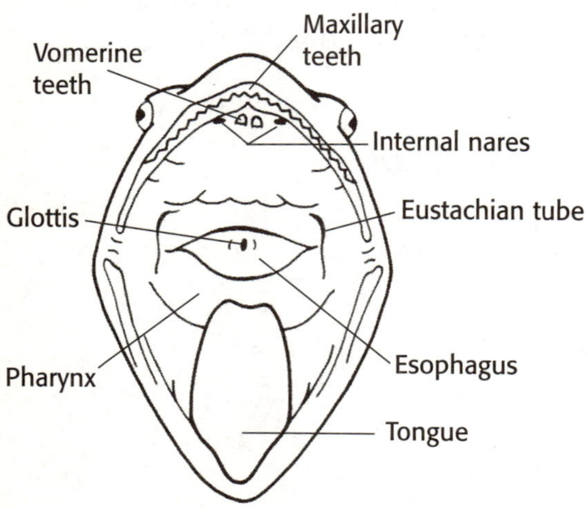

FIGURE 4 INCISIONS FOR DISSECTION

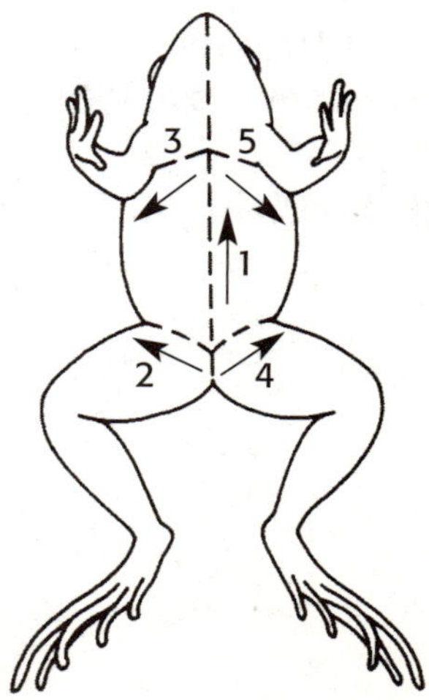

8. Lift and cut through the muscles and breastbone to open up the body cavity. If your frog is a female, the abdominal cavity may be filled with dark-colored eggs. If so, remove the eggs on one side so you can see the organs beneath them.

9. Review the following information about the frog's digestive system.

The digestive system consists of the organs of the digestive tract, or food tube, and the digestive glands. From the esophagus, swallowed food moves into the *stomach* and then into the *small intestine.* Bile is a digestive juice made by the *liver* and stored in the *gallbladder.* Bile flows into a tube called the *common bile duct,* into which *pancreatic juice,* a digestive juice from the *pancreas,* also flows. The contents of the common bile duct flow into the small intestine, where most of the digestion and absorption of food into the bloodstream takes place. Indigestible materials pass through the *large intestine* and then into the *cloaca*—the common exit chamber of the digestive, excretory, and reproductive systems.

10. Use **Figure 5** to locate and identify the organs of the digestive system: esophagus, stomach, small intestine, large intestine, cloaca, liver, gallbladder, and pancreas.

FIGURE 5 ABDOMINAL AND CHEST CAVITIES OF FROG

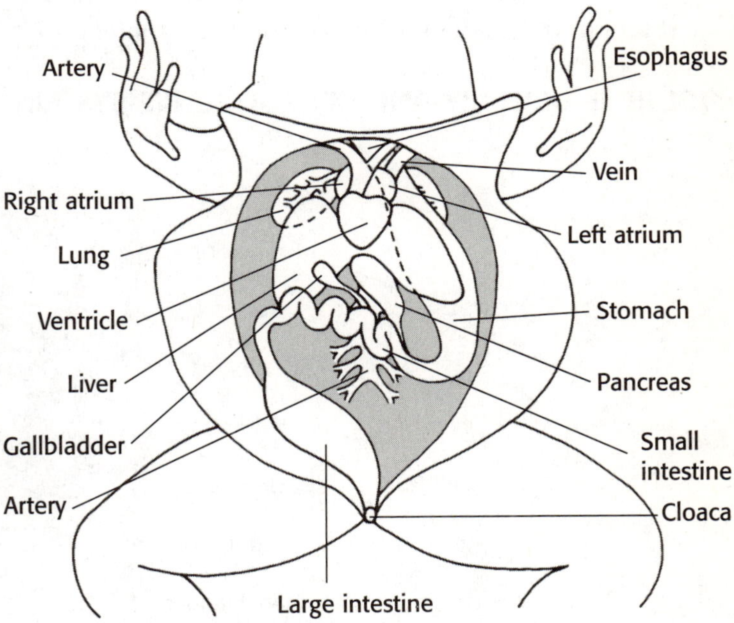

11. Review the following information about the frog's respiratory and circulatory systems.

The respiratory system consists of the nostrils and the larynx, which opens into two *lungs*—hollow sacs with thin walls. The walls of the lungs are filled with *capillaries,* which are microscopic blood vessels through which materials pass into and out of the blood. The circulatory system consists of the heart, blood vessels, and blood. The heart has two receiving chambers, or *atria,* and one sending chamber, or *ventricle.* Blood is carried to the heart in vessels called *veins.* Veins from different parts of the body enter the right and left atria. Blood from both atria goes into the ventricle and then is pumped into the *arteries,* which are blood vessels that carry blood away from the heart.

12. Use **Figure 5** to locate and identify the left atrium, right atrium, and ventricle of the heart. Find an artery attached to the heart and another artery near the backbone. Find a vein near one of the shoulders. Find the two lungs.

13. Review the following information about the frog's urinary and reproductive systems.

The urinary system consists of the frog's kidneys, ureters, bladder, and cloaca. The *kidneys* are organs that excrete urine. Connected to each kidney is a *ureter*—a tube through which urine passes into the *urinary bladder*. This is a sac that stores urine until it passes out of the body through the cloaca.

The organs of the male reproductive system are the testes, sperm ducts, and cloaca. Those of the female system are the ovaries, oviducts, uteri, and cloaca. The *testes* produce *sperm*, or male sex cells, which move through *sperm ducts*—tubes that carry sperm into the cloaca, from which the sperm move outside the body. The *ovaries* produce *eggs*, or female sex cells, which move through *oviducts* into the *uteri*, then through the cloaca outside the body.

14. Use a probe and scissors to lift and remove the intestines and liver. Use **Figure 6** to identify the parts of the urinary and reproductive systems.

15. Remove the peritoneal membrane, which is connective tissue that lies on top of the red kidneys. Observe the yellow fat bodies that are attached to the kidneys. Find the ureters and the urinary bladder. Find the testes and sperm ducts in the male and the ovaries, oviducts, and uteri in the female.

FIGURE 6 URINARY AND REPRODUCTIVE SYSTEMS OF THE FROG

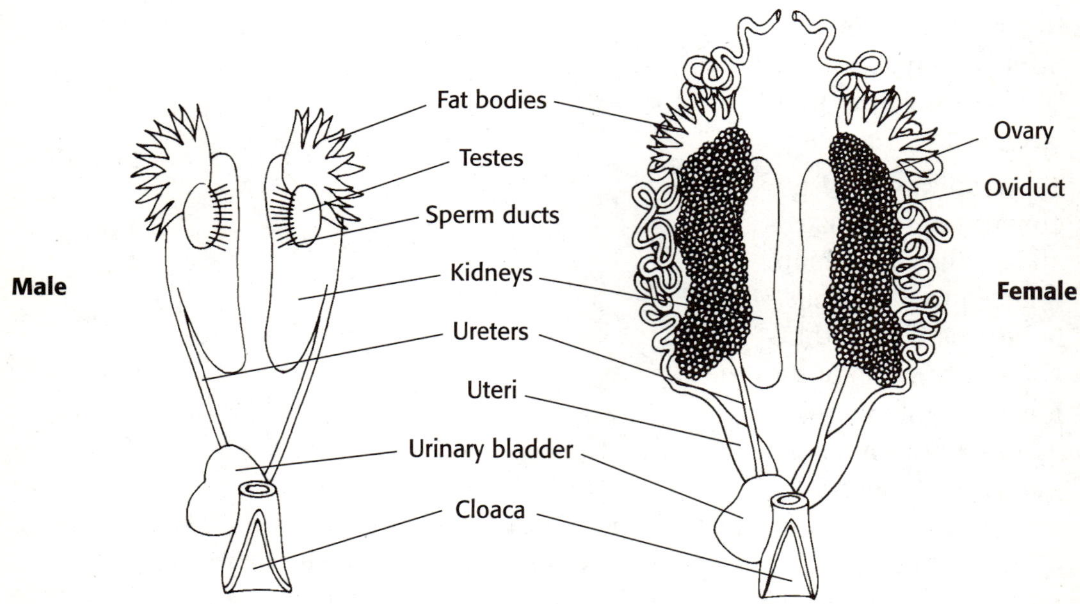

16. Remove the kidneys and look for threadlike spinal nerves that extend from the spinal cord. Dissect a thigh, and trace one nerve into a leg muscle. Note the size and texture of the leg muscles.

17. Dispose of your materials according to your teacher's instructions.

18. Clean up your work area, and wash your hands thoroughly before leaving the lab.

Analysis

1. **Identifying Relationships** What do you think is the function of the nictitating membrane? Why?

____

____

____

2. **Analyzing Data** A frog does not chew its food. What do the positions of its teeth suggest about how the frog uses them?

____

____

3. **Organizing Data** Through which structures does food pass as it moves through a frog's digestive tract?

____

____

____

4. **Organizing Data** List the structures, in order, through which sperm pass in a male and eggs pass in a female.

____

____

5. **Organizing Data** Through which structures does urine pass in a frog?

____

____

6. **Examining Data** Which parts of the frog's nervous system can be observed in its abdominal cavity and hind leg?

____

____

7. **Making Inferences** The abdominal cavity of a frog at the end of the hibernation season would contain very small fat bodies or none at all. What is the function of the fat bodies?

____

____

8. Analyzing Data What features did you notice in the frog that have helped it adapt to living on land?

____

____

9. Explaining Events Why did you dissect the nerves last?

____

____

Conclusions

1. Making Predictions Structures of an animal's body that help it survive in its environment are adaptations. How do the frog's powerful hind legs help it survive both in water and on land?

____

____

2. Applying Conclusions Suppose in a living frog the spinal nerve extending to the leg muscle was cut. What ability would the frog lose? Why?

____

____

____

3. Drawing Conclusions Think about the skin of the frog you dissected. A frog can respire through its skin, but a fish cannot. Why do you think this is so?

____

____

____

Extensions

1. Designing Experiments Before disposing of your frog, expose the brain and spinal cord, under your teacher's guidance. Use a textbook illustration to identify the parts of the brain and the spinal cord.

2. Research and Communications Investigate the decline of populations of frogs in some areas of the United States. Discuss the reported increase in frog deformities.

Name ______________________ Class ______________ Date ______________

Data Lab

DATASHEET FOR IN-TEXT LAB

Analyzing Ion Excretion in Fish

Background

A few species of fish, such as adult salmon, are able to move between saltwater and freshwater environments. The graph below shows the excreted ion concentration of a fish as it travels from one body of water to another. Examine the graph, and answer the analysis questions.

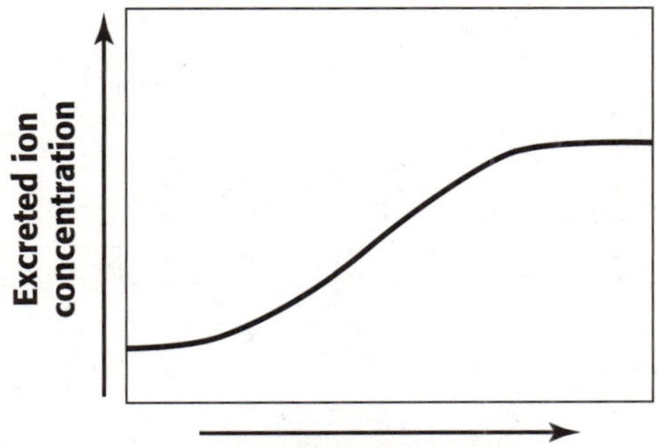

Analysis

1. **Determine** if the fish is losing or gaining ions by excretion as it travels.

   losing ions by excretion

2. **Critical Thinking**
   Inferring Conclusions Is the fish traveling from fresh water to salt water or from salt water to fresh water?

   from fresh water to salt water

3. **Summarize** the reasoning you used to answer item 2.

   The graph shows that the farther the fish travels, the greater is its excreted ion concentration. A fish's tissues have a lower concentration of ions than is found in salt water, and they lose water through osmosis. One would expect the fish to excrete more ions as the water becomes saltier.

Name ______________________ Class ______________ Date ____________

Quick Lab

DATASHEET FOR IN-TEXT LAB

Modeling the Action of a Swim Bladder

Most fish use a swim bladder to regulate their depth in water. As gas enters the swim bladder, the fish rises in the water. As gas is expelled, the fish sinks to a lower depth.

MATERIALS

- 100 mL beaker or small glass
- cold, clear, carbonated soft drink
- 2 very dry raisins

Procedure

1. Fill a 100 mL beaker with a cold, carbonated soft drink.
2. Drop two raisins into the beaker, and observe what happens over the next 5 minutes.

Analysis

1. **Describe** what happened after you dropped the raisins into the soft drink.

   The raisins sank to the bottom. As bubbles collected on them, the raisins rose to the top. When they reached the top, they sank again.

2. **Forming Hypotheses** Develop a hypothesis to explain your observations.

   The bubbles clung to the raisins and imparted buoyancy, causing the raisins to rise. When the raisins reached the top, the bubbles burst, and the raisins sank again.

3. **Critical Thinking**
   Analyzing Results How does the lifting of the raisins differ from the use of a swim bladder to control buoyancy?

   The bubbles formed on the outside of the raisin. A swim bladder is internal. Also, the amount of gas in the swim bladder can be increased or decreased, unlike the bubbles on the raisins.

Name ______________________ Class ______________ Date ______________

Modeling the Action of a Swim Bladder *continued*

4. Critical Thinking

Forming Reasoned Opinions Think about the energy you would have to expend to keep yourself in one position underwater. What advantage might a swim bladder provide to a fish?

It would take a great deal of energy to stay in one position underwater with two lungs full of air. A swim bladder offers a fish tremendous savings of energy because the fish controls the amount of gas in the swim bladder. This makes more energy available for daily activities, such as catching prey or escaping predators.

Name ______________________ Class ______________ Date ____________

Exploration Lab

DATASHEET FOR IN-TEXT LAB

Observing a Live Frog

SKILLS

- Observing
- Relating

OBJECTIVES

- **Examine** the external features of a frog.
- **Observe** the behavior of a frog.
- **Explain** how a frog is adapted to life on land and in water.

MATERIALS

- live frog in a terrarium
- live insects (crickets or mealworms)
- 600 mL beaker
- aquarium half-filled with dechlorinated water

Before You Begin

Frogs, which are **amphibians,** are adapted for living on land and in water. For example, a frog's eyes have an extra eyelid called the **nictitating membrane.** This eyelid protects the eye when the frog is underwater and keeps the eye moist when the frog is on land. The smooth skin of a frog acts as a respiratory organ by exchanging oxygen and carbon dioxide with the air or water. The limbs of a frog enable it to move both on land and in water. In this lab, you will examine a live frog in both a terrestrial environment and an aquatic environment.

1. Write a definition for each boldface term in the paragraph above and for the following term: tympanic membrane. Use a separate sheet of paper.
   Answers appear in the TE for this lab.
2. You will be using the data table provided to record your data.
3. Based on the objectives for this lab, write a question you would like to explore about frogs.
   Answers will vary. For example: How are frogs adapted for living on land and in water?

Procedure

PART A: OBSERVING A FROG

1. Observe a live frog in a terrarium. Closely examine the external features of the frog. Make a drawing of the frog. Label the eyes, nostrils, tympanic membranes, front legs, and hind legs.

2. Watch the frog's movements as it breathes air into and out of its lungs. Record your observations.

Data Table	
Behavior/structure	**Observations**
Breathing	
Eyes	
Legs	
Response to food	
Response to noise	
Skin	
Swimming behavior	

3. Look closely at the frog's eyes, and note their location. Examine the upper and lower eyelids as well as a third transparent eyelid called a *nictitating membrane*. Describe how the eyelids move.

4. Study the frog's legs, and note the difference between the front and hind legs.

5. Place a live insect, such as a cricket or a mealworm, into the terrarium. Observe how the frog reacts.

6. Tap the side of the terrarium farthest from the frog, and observe the frog's response.

7. Place a 600 mL beaker in the terrarium. **CAUTION: Handle live frogs gently. Frogs are slippery! Do not allow a frog to injure itself by jumping from a lab table to the floor.** Carefully pick up the frog, and examine its skin. How does it feel? Now place the frog in the beaker. Cover the beaker with your hand, and carry it to a freshwater aquarium. Tilt the beaker, and gently lower it into the water until the frog swims out.

8. Watch the frog float and swim. Notice how the frog uses its legs to swim. Also notice the position of the frog's head. As the frog swims, bend down to view the underside of the frog. Then look down on the frog from above. Compare the color on the dorsal and ventral sides of the frog.

PART B: CLEANUP AND DISPOSAL

9. Dispose of broken glass in the designated waste containers. Put live animals in the designated containers. Do not pour chemicals down the drain or put lab materials in the trash unless your teacher tells you to do so.

10. Clean up your work area and all lab equipment. Return lab equipment to its proper place. Wash your hands thoroughly before you leave the lab and after you finish all work.

Name ______________________ Class ______________ Date ______________

Analyze and Conclude

1. **Summarizing Information** How does a frog use its hind legs for moving on land and in water?

   A frog uses its long hind legs for jumping on land and for swimming in water.

2. **Recognizing Relationships** How does the position of a frog's eyes benefit the frog while it is swimming?

   Having eyes on the top of its head enables a frog to see above the waterline while the rest of the frog is submerged.

3. **Analyzing Data** What features of an adult frog provide evidence that it has an aquatic life and a terrestrial life?

   aquatic life–webbed feet, eyes on top of head, respires through skin; terrestrial life–jumping legs, breathes with lungs

4. **Analyzing Methods** Were you able to determine in this lab how a frog hears? Explain.

   Answers will vary. Students should recognize that although they observed the tympanic membrane, they cannot determine how a frog hears by observing its actions.

5. **Inferring Conclusions** What can you infer about a frog's field of vision from the position of its eyes?

   A frog can see in almost all directions.

6. **Forming Hypotheses** How is the coloration on the dorsal and ventral sides of a frog an adaptive advantage?

   As viewed from above, the dark dorsal skin blends in with the color of the water, making the frog less visible to predators from above. Viewed from below, the light ventral skin blends in with the color of the sky, making the frog less visible to aquatic predators.

7. **Further Inquiry** Write a new question about frogs that could be explored with another investigation.

   Answers will vary. For example: How do two frogs interact on land and in water?

Exploration Lab

FIELD ACTIVITY

Observing Structure and Behavior in Fishes

Teacher Notes

TIME REQUIRED One 45-minute period

SKILLS ACQUIRED

Collecting data
Identifying and recognizing patterns
Inferring
Interpreting
Predicting

RATINGS Easy ← 1 2 3 4 → Hard

Teacher Prep–1
Student Setup–1
Concept Level–2
Cleanup–1

THE SCIENTIFIC METHOD

Make Observations Students observe body shapes and behaviors of various species of fish in an aquarium.

Analyze the Results Analysis questions 1 and 2 require students to analyze their observations.

Draw Conclusions Conclusions question 1 asks students to draw conclusions about the survival value of camouflage.

TIPS AND TRICKS

Preparation

Contact the staff of the aquarium you will visit ahead of time. Inform them of the date of your visit, the number of students in the class, and the observations that students are expected to make. The staff may be able to recommend specific tanks that students should include in their observations.

Procedure

Other fishes with a streamlined body include anchovy, dace, bonito, mackerel, and most sharks. Other fishes with a flattened body include halibut and turbot.

If students are having difficulty deciding where to look for camouflage and feeding behavior, use the sample answers in Tables 2 and 3 for suggestions to get students started. A feeding behavior not listed in Table 3 is angling, in which some bottom-dwelling fish wiggle a fleshy tentacle in front of their mouth to lure prey. This behavior is exhibited by the anglerfish, goosefish, batfish, frogfish, and sea toad.

If students cannot observe fish feeding, they may be able to learn about the feeding methods of certain species by reading interpretive signs near the tanks.

Observing Structure and Behavior in Fishes *continued*

For further information on fish behavior, students can consult references such as *The World of Fish*, by Franco De Carli, Abbeville Press, 1975, a volume of the *Abbeville Press Encyclopedia of Natural Science.* This compact book has excellent color illustrations and discussions of fish behavior.

Students who wish to complete Extensions item 1 can use fish such as scissortails *(Rasbora trilinea)*, rosy tetras *(Hyphesobrycea rosaceus)*, tiger barbs *(Barbus tetrazona)*, zebra fish *(Brachydanio rerio)*, or minnows (many species). Have students consult a reference on tropical fish for advice on conditioning the aquarium water.

Name ______________________ Class ______________ Date ______________

Exploration Lab **FIELD ACTIVITY**

Observing Structure and Behavior in Fishes

Most fishes share several characteristics, including gills, fins, and scales, that distinguish them from other vertebrates. Beyond these shared characteristics, the 25,000 species of fishes that inhabit Earth's oceans, lakes, ponds, and rivers exhibit a great diversity of body styles and behaviors.

In this lab, you will observe the structural and behavioral adaptations of several species of fish at an aquarium. You will then consider how these adaptations help each species survive in its habitat.

OBJECTIVES

Observe body shapes, camouflage, feeding behavior, and schooling behavior in various species of fish.

Relate observed adaptations to the survival of each species.

MATERIALS

- clipboard and paper or notebook
- pencil or pen

Procedure

BODY SHAPES

1. Find a tank that contains trout, herring, barracuda, or tuna. If the aquarium does not have any of these fishes, ask your teacher for some alternatives.
2. Identify the species of fish in the tank. In **Table 1,** record the common and scientific names of the species.
3. Describe the shape of the fish's body. In **Table 1,** record your description.
4. Observe the activity of the fish. How does it spend most of its time? In **Table 1,** record your observations.
5. Repeat steps 2–4 with another species of fish from each of the following two groups: (1) flounder, sole, ray, or skate; and (2) eel or lamprey.

TABLE 1 BODY SHAPES AND ACTIVITY

Species	Body shape	Activity
Trout, herring, barracuda, or tuna	**sleek, streamlined, spindle-shaped**	**rapid swimming in open water**
Flounder, sole, ray, or skate	**flattened from top to bottom**	**resting or creeping along bottom of tank**
Eel or lamprey	**elongated, snakelike**	**attached to rock or hiding in borrow or crevice**

Sample observations provided above.

Name ______________________ Class ______________ Date ____________

Observing Structure and Behavior in Fishes *continued*

CAMOUFLAGE

6. Find three examples of fish that use body colors, markings, or postures to blend with their surroundings.

7. For each species, record its common and scientific names in **Table 2,** and describe the type of camouflage it uses.

TABLE 2 CAMOUFLAGE

Species	Type of camouflage
Example: many swimming fishes	**are dark on top, which makes them hard to see from above, and are light on the bottom, which makes them hard to see from below**
Example: flatfish	**changes the color of its top surface to match the color of the ground where it is lying**
Example: pipefish	**slender, elongated body oriented vertically among blades of sea grasses**

FEEDING BEHAVIOR

8. Find three examples of fish that use different methods for feeding.

9. For each species, record its common and scientific names in **Table 3,** and describe its feeding method.

TABLE 3 FEEDING BEHAVIOR

Species	Type of feeding behavior
Example: shark	**uses sharp teeth to tear off pieces of flesh**
Example: sea horse	**sucks water and small prey into its mouth**
Example: many herringlike fishes	**strain tiny organisms from the water that passes through their gills**

SCHOOLING BEHAVIOR

10. Look for a tank that contains a large group of fish. Decide whether or not the group is a school. Among the kinds of fish that form schools are anchovy, barracuda, bluefish, bonito, dace, herring, mackerel, and shark.

What criteria would help you decide whether a group of fish is a school?

If there are many fish in the group and they move in unison, the group is likely to be a school.

11. Identify the species of fish in the school. In **Table 4,** record the common and scientific names of the species.

Name ______________________ Class ______________ Date ______________

Observing Structure and Behavior in Fishes *continued*

TABLE 4 SCHOOLING BEHAVIOR

Species	Number of individuals	Observations
Answers will vary.		**Answers will vary but should refer to whether or how the school breaks up and reforms.**

12. Count or estimate the number of individuals in the school. Record the number in **Table 4.**

13. Observe the school carefully. Do the fish spend all of their time in the school, or does the school break up occasionally and then reform? If it breaks up and reforms, what appears to trigger school formation? Record your observations in **Table 4.**

14. Repeat steps 11–13 with another school of fish in a different tank.

Analysis

1. Identifying Relationships How are the body shapes of the fishes you observed related to the types of activities that those fishes engage in?

Answers will vary. Fish that swim rapidly in open water typically have streamlined bodies. Fish that creep along the bottom often have flattened bodies. Fish that hide in burrows or crevices may have elongated, snakelike bodies.

2. Recognizing Patterns What structural adaptations are associated with the different feeding behaviors that you observed?

Answers will vary. Fish that tear off pieces of flesh usually have sharp teeth. Fish that strain tiny organisms from the water have sievelike structures in their mouths or near their gills.

Name ______________________ Class ______________ Date ______________

Conclusions

1. **Drawing Conclusions** How does camouflage help fish to survive in the cases where you observed it?

   Answers will vary. Different kinds of fish may use camouflage to hide from predators, prey, or both.

2. **Applying Conclusions** Would you expect camouflage to be more important for fish that live in shallow water or for those that live in very deep water? Explain your reasoning.

   Camouflage should be more important for fish that live in shallow water because enough light penetrates the water to make a fish's color and markings visible. In very deep water, it is always dark, so camouflage is not useful.

3. **Making Predictions** Why might it be advantageous for fish to swim in schools?

   Answers will vary. An individual fish may be less likely to be preyed on when it is part of a school than when it is alone. Swimming in schools may help fish find scarce food resources and mates. Fish may swim more efficiently when large numbers of them swim close together in a school.

Extensions

1. **Designing Experiments** Think of a question you would like to answer about fish structure or behavior. Design an experiment that could answer your question, using fish sold at a local pet store and materials available in your classroom.

2. **Research and Communications** Research another class of vertebrates in which individuals move in groups. Compare their movements with those of the fish you observed. Report the results of your research in an oral presentation.

Skills Practice Lab

OBSERVATION

Frog Dissection

Teacher Notes

TIME REQUIRED One or two 45-minute periods

SKILLS ACQUIRED

Collecting data
Identifying patterns
Inferring
Interpreting
Organizing and analyzing data

RATING Easy ← 1 2 3 4 → Hard

Teacher Prep–2
Student Setup–2
Concept Level–2
Cleanup–2

THE SCIENTIFIC METHOD

Make Observations Students make observations about frog anatomy as they dissect a frog.

Analyze the Results Analysis questions 1, 2, 6, 7, and 8 require students to analyze their results.

Draw Conclusions Conclusions questions 1–3 require students to draw conclusions from their data.

MATERIALS

Materials for this lab can be purchased from WARD'S. See the *Master Materials List* for ordering instructions.

SAFETY CAUTIONS

- Discuss all safety symbols with students.
- Do not use specimens preserved with formaldehyde.
- Prolonged contact with WARDSafe, which should be used to store the specimens, may be irritating to skin and eyes and may cause allergic reaction in hypersensitive individuals. Discontinue use if redness or swelling occurs. In case of contact, flush with water, including under eyelids, for 15 minutes. Contact physician if irritation or redness persists. May be toxic if swallowed. If the student is conscious, drink 8–10 oz (240–300 mL) water to dilute material; Induce vomiting. Get prompt medical attention.
- Carefully instruct students in the correct handling of all dissecting instruments, especially the scalpel. Do not allow students to use razor blades.

DISPOSAL

Wrap the remains of dissected frogs in old newspaper, place in a plastic "biohazard" bag, and tie the bag securely. Inform the school custodian of the bag's contents, and hand it over personally for its safe disposal.

TECHNIQUES TO DEMONSTRATE

Before the lab, demonstrate the use of the dissection kit including the probe, forceps, and scalpel as well as how to pin the specimen. Show students how to hold the scalpel properly and how to use short strokes to make an incision.

TIPS AND TRICKS

Preparation

This lab works best in groups of two students.

All WARD'S specimens are treated with WARDSafe. These specimens require no additional preparation. They greatly minimize student discomfort, especially eye irritation.

Procedure

Emphasize to students that dissection is not haphazard "cutting and slicing." Rather, it is making careful incisions to expose parts, and then using a probe to separate organs from their coverings. Point out that the intent is to carefully unwrap the animal's structures without damaging the structures.

Between class periods, have students cover each specimen pinned to the dissection tray with a paper towel dampened with WARDSafe. Each group should then store their specimen in a plastic bag that is tied securely. Have students use a felt-tip marker to write their names directly on the plastic bags.

If you do not wish to have students perform this dissection, alternative methods of studying the structures are available. These include models, filmstrips, videocassettes, and computer simulations.

Give the following instructions to students who carry out Extensions question 1. Cut off the tip of the nose and gently pick and scrape away the top of the skull to expose the brain. Then carefully dissect part of the spinal cord. Have everyone in the class observe the completed dissection and use a textbook illustration to identify the olfactory lobes, cerebral hemisphere, optic lobes, cerebellum, cranial nerves, medulla, spinal cord, and spinal nerves. Discuss the functions of each part.

Name ______________________________ Class ________________ Date ______________

Skills Practice Lab **OBSERVATION**

Frog Dissection

Amphibians, such as frogs, are examples of animals that have structural adaptations for living in or near water and for living on land. As tadpoles, they have gills, but adult frogs have lungs. Tadpoles have tails, but adult frogs have legs. Many changes occur in the metamorphosis of a tadpole. The structures that frogs have depend on the stage of the frog's life cycle. In this lab, you will dissect a frog in order to observe the external and internal structures of frog anatomy.

MATERIALS

- dissecting pins (6–10)
- dissecting tray
- dull probe or dissecting needle
- felt-tip marker
- forceps
- gloves
- lab apron
- paper towels
- plastic storage bag and twist tie
- preserved frog
- safety goggles
- scissors

OBJECTIVES

Describe the appearance of various organs found in a frog.

Name the organs that make up various systems of a frog.

Procedure

1. Put on safety goggles, gloves, and a lab apron.
2. Rinse and place a frog on a dissection tray with its dorsal side down. To determine the frog's sex, look at the hand digits, or fingers, on its forelegs. A male frog usually has thick pads on its "thumbs," which is one external difference between the sexes, as shown in **Figure 1.** Male frogs are also usually smaller than female frogs. Observe several frogs to see the difference between males and females.

FIGURE 1 HAND DIGITS OF MALE AND FEMALE FROG

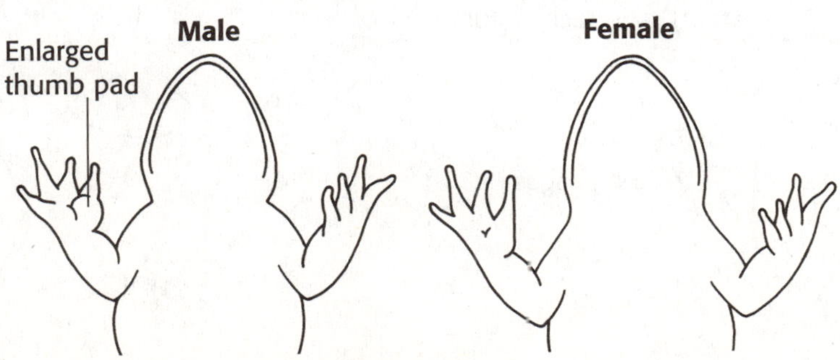

3. Review the following information about the frog's head.

On the outside of the frog's head are two *external nares*, or nostrils; two *tympani*, or eardrums; and two eyes, each of which has three lids. The third lid, called the *nictitating membrane*, is transparent.

4. Use **Figure 2** to locate and identify the external features of the head. Find the mouth, external nares, tympani, eyes, and nictitating membranes.

5. Review the following information about the frog's mouth.

Inside the mouth are two *internal nares*, or openings into the nostrils; two *vomerine teeth* in the middle of the roof of the mouth; and two *maxillary teeth* at the sides of the mouth. Also inside the mouth behind the tongue is the *pharynx*, or throat. In the pharynx, there are several openings: one into the *esophagus*, the tube into which food is swallowed; one into the *glottis*, through which air enters the *larynx*, or voice box; and two into the *Eustachian tubes*, which connect the pharynx to the ear.

FIGURE 2 EXTERNAL FEATURES OF FROG HEAD

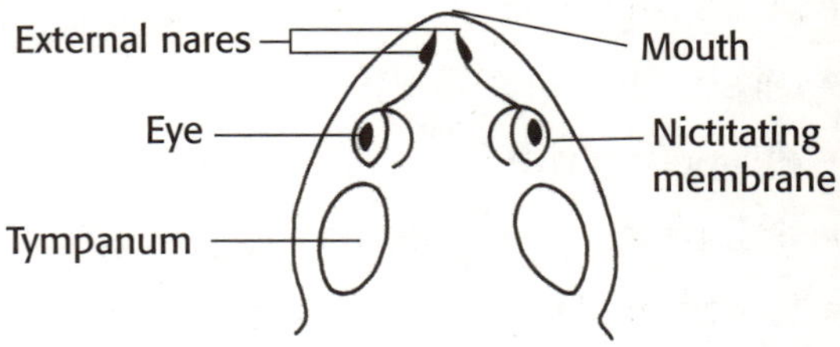

6. Turn the frog on its back, and pin down the legs. Cut the hinges of the mouth, and open it wide. Use **Figure 3** to locate and identify the structures inside the mouth. Use a dull probe to help find each part: the vomerine teeth, the maxillary teeth, the internal nares, the tongue, the openings to the Eustachian tubes, the esophagus, the pharynx, and the slitlike glottis.

7. Look for the opening to the frog's cloaca, located between the hind legs. Use forceps to lift the skin, and use scissors to cut along the center of the body from the cloaca to the lip. Turn back the skin, cut toward the side at each leg, and pin the skin flat. **Figure 4** shows how to make these cuts.

FIGURE 3 STRUCTURES OF FROG MOUTH

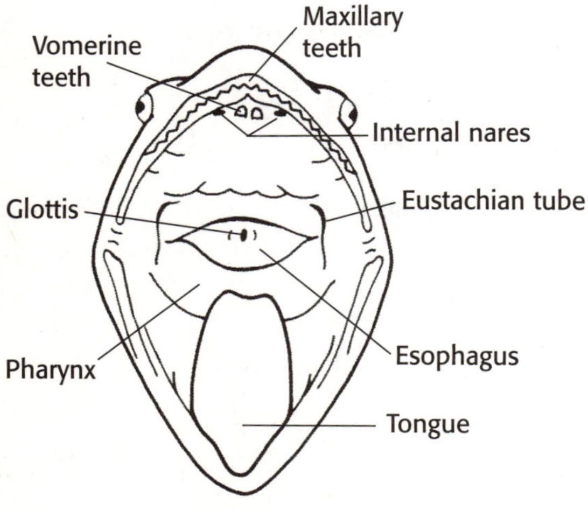

FIGURE 4 INCISIONS FOR DISSECTION

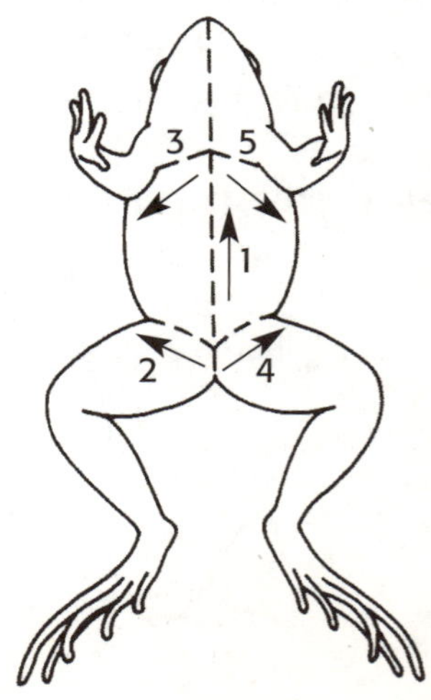

Name ______________________ Class ______________ Date ____________

Frog Dissection *continued*

8. Lift and cut through the muscles and breastbone to open up the body cavity. If your frog is a female, the abdominal cavity may be filled with dark-colored eggs. If so, remove the eggs on one side so you can see the organs beneath them.

9. Review the following information about the frog's digestive system.

The digestive system consists of the organs of the digestive tract, or food tube, and the digestive glands. From the esophagus, swallowed food moves into the *stomach* and then into the *small intestine.* Bile is a digestive juice made by the *liver* and stored in the *gallbladder.* Bile flows into a tube called the *common bile duct*, into which *pancreatic juice*, a digestive juice from the *pancreas*, also flows. The contents of the common bile duct flow into the small intestine, where most of the digestion and absorption of food into the bloodstream takes place. Indigestible materials pass through the *large intestine* and then into the *cloaca*—the common exit chamber of the digestive, excretory, and reproductive systems.

10. Use **Figure 5** to locate and identify the organs of the digestive system: esophagus, stomach, small intestine, large intestine, cloaca, liver, gallbladder, and pancreas.

FIGURE 5 ABDOMINAL AND CHEST CAVITIES OF FROG

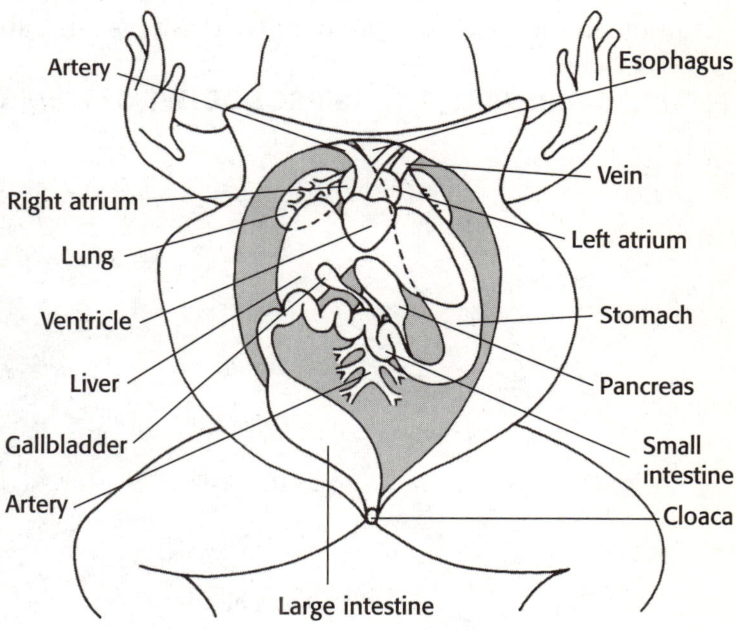

11. Review the following information about the frog's respiratory and circulatory systems.

The respiratory system consists of the nostrils and the larynx, which opens into two *lungs*—hollow sacs with thin walls. The walls of the lungs are filled with *capillaries*, which are microscopic blood vessels through which materials pass into and out of the blood. The circulatory system consists of the heart, blood vessels, and blood. The heart has two receiving chambers, or *atria*, and one sending chamber, or *ventricle.* Blood is carried to the heart in vessels called *veins.* Veins from different parts of the body enter the right and left atria. Blood from both atria goes into the ventricle and then is pumped into the *arteries*, which are blood vessels that carry blood away from the heart.

12. Use **Figure 5** to locate and identify the left atrium, right atrium, and ventricle of the heart. Find an artery attached to the heart and another artery near the backbone. Find a vein near one of the shoulders. Find the two lungs.

Name ______________________ Class ______________ Date ____________

13. Review the following information about the frog's urinary and reproductive systems.

The urinary system consists of the frog's kidneys, ureters, bladder, and cloaca. The *kidneys* are organs that excrete urine. Connected to each kidney is a *ureter*—a tube through which urine passes into the *urinary bladder.* This is a sac that stores urine until it passes out of the body through the cloaca.

The organs of the male reproductive system are the testes, sperm ducts, and cloaca. Those of the female system are the ovaries, oviducts, uteri, and cloaca. The *testes* produce *sperm*, or male sex cells, which move through *sperm ducts*—tubes that carry sperm into the cloaca, from which the sperm move outside the body. The *ovaries* produce *eggs*, or female sex cells, which move through *oviducts* into the *uteri*, then through the cloaca outside the body.

14. Use a probe and scissors to lift and remove the intestines and liver. Use **Figure 6** to identify the parts of the urinary and reproductive systems.

15. Remove the peritoneal membrane, which is connective tissue that lies on top of the red kidneys. Observe the yellow fat bodies that are attached to the kidneys. Find the ureters and the urinary bladder. Find the testes and sperm ducts in the male and the ovaries, oviducts, and uteri in the female.

FIGURE 6 URINARY AND REPRODUCTIVE SYSTEMS OF THE FROG

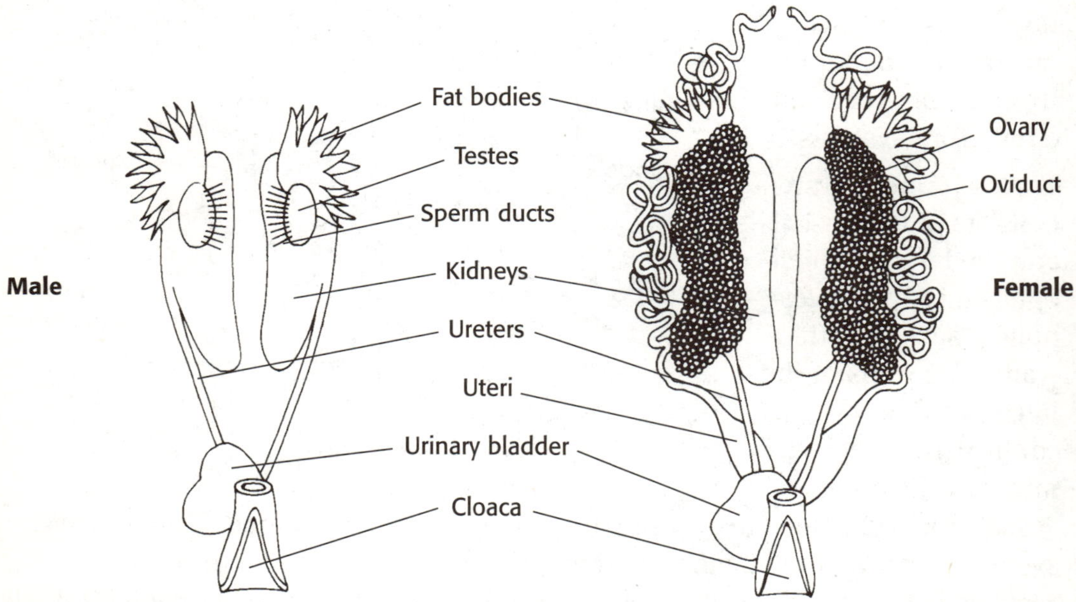

16. Remove the kidneys and look for threadlike spinal nerves that extend from the spinal cord. Dissect a thigh, and trace one nerve into a leg muscle. Note the size and texture of the leg muscles.

17. Dispose of your materials according to your teacher's instructions.

18. Clean up your work area, and wash your hands thoroughly before leaving the lab.

Analysis

1. **Identifying Relationships** What do you think is the function of the nictitating membrane? Why?

   Because the membrane is transparent, the nictitating membrane may permit the frog to see underwater while protecting the eye.

2. **Analyzing Data** A frog does not chew its food. What do the positions of its teeth suggest about how the frog uses them?

   The teeth hold onto living prey until it can be swallowed.

3. **Organizing Data** Through which structures does food pass as it moves through a frog's digestive tract?

   Food passes through the esophagus, stomach, small intestine, large intestine, and cloaca.

4. **Organizing Data** List the structures, in order, through which sperm pass in a male and eggs pass in a female.

   Sperm pass through the testis, sperm duct, and cloaca. Eggs pass through the ovaries, oviducts, uteri, and cloaca.

5. **Organizing Data** Through which structures does urine pass in a frog?

   Urine passes through the kidneys, ureters, urinary bladder, and cloaca.

6. **Examining Data** Which parts of the frog's nervous system can be observed in its abdominal cavity and hind leg?

   The spinal cord (in the backbone) and spinal nerves can be observed in the abdominal cavity and hind leg.

7. **Making Inferences** The abdominal cavity of a frog at the end of the hibernation season would contain very small fat bodies or none at all. What is the function of the fat bodies?

   They are a source of stored food during hibernation.

8. **Analyzing Data** What features did you notice in the frog that have helped it adapt to living on land?

   Legs, lungs, and a heart with two atria and a ventricle helped frogs adapt to life on land.

9. **Explaining Events** Why did you dissect the nerves last?

   Other structures covered the nerves and had to be dissected first.

Conclusions

1. **Making Predictions** Structures of an animal's body that help it survive in its environment are adaptations. How do the frog's powerful hind legs help it survive both in water and on land?

   The legs enable the frog to both swim in water and jump on land and thus catch food and escape predators.

2. **Applying Conclusions** Suppose in a living frog the spinal nerve extending to the leg muscle was cut. What ability would the frog lose? Why?

   The frog would lose the ability to swim or jump. The spinal nerves conduct impulses from the spinal cord and brain, which control muscle movement.

3. **Drawing Conclusions** Think about the skin of the frog you dissected. A frog can respire through its skin, but a fish cannot. Why do you think this is so?

   The scales that cover the skin of a fish do not allow oxygen molecules to diffuse across the fish's skin. A frog's skin is thin and moist, allowing diffusion of oxygen.

Extensions

1. **Designing Experiments** Before disposing of your frog, expose the brain and spinal cord, under your teacher's guidance. Use a textbook illustration to identify the parts of the brain and the spinal cord.
2. **Research and Communications** Investigate the decline of populations of frogs in some areas of the United States. Discuss the reported increase in frog deformities.

Answer Key

Directed Reading

SECTION: THE FISH BODY

1. the oxygen dissolved in water
2. heart
3. cartilage
4. externally
5. spawning
6. inside
7. gill filaments
8. gill slit
9. countercurrent flow
10. b
11. d
12. a
13. c
14. Sea water has a much higher salt concentration than the cells of marine fishes, so marine fishes continually lose water by osmosis. They must drink sea water and actively pump excess ions out of their body. Freshwater fishes continually take in water by osmosis and regain salts by actively taking them in from their environment.
15. Nephrons are units of the kidneys that regulate the body's salt and water balance and remove metabolic wastes from the blood.

SECTION: TODAY'S FISHES

1. operculum
2. lateral line
3. swim bladder
4. Lampreys and hagfishes are jawless and have scaleless, eel-like bodies with multiple gill slits and unpaired fins. Their skeletons are made of cartilage, and they retain their notochord into adulthood.
5. Sharks are streamlined, and their cartilage is reinforced with calcium carbonate, making the cartilage very light and strong.
6. Rays and skates have flattened bodies. Most species also have flattened teeth, which are used to crush small fishes and invertebrates.
7. Opercula enable bony fishes to move water over their gills while remaining stationary in the water, thereby conserving energy.
8. The lateral line enables bony fishes to perceive their position and rate of movement and to detect motionless objects by sensing the movement of water reflected by the objects.
9. Gas enters the swim bladder from the bloodstream.
10. Teleosts are bony fishes with highly mobile fins, very thin scales, and completely symmetrical tails.
11. Lobe-finned bony fishes are bony fishes with paired fins that usually consist of a long, fleshy, muscular lobe supported by a central core of bones.

SECTION: AMPHIBIANS

1. frogs and toads (Anura), salamanders and newts (Urodela), and caecilians (Apoda)
2. legs, lungs, double-loop circulation, a partially divided heart, and cutaneous respiration
3. Pulmonary veins enable the heart to pump oxygen-rich blood to the tissues at a higher pressure, which increases the rate of oxygen delivery to body tissues.
4. The septum helps keep oxygen-rich blood from mixing with oxygen-poor blood as each enters the heart.
5. pulmonary veins
6. septum
7. lung
8. shorter
9. smooth
10. is transformed from a tadpole
11. external
12. cloaca
13. moist places
14. amphibians

Active Reading

SECTION: THE FISH BODY

1. to force the water from the mouth, over the gills, and out through the gill slits

2. A gill is made up of rows of gill filaments.
3. Countercurrent flow ensures that oxygen diffuses into the blood over the entire length of the capillaries in the gills.
4. c

SECTION: TODAY'S FISHES

1. a lateral line system, a gill cover, and a swim bladder
2. a hard plate that covers the gills of most bony fishes
3. It helps a bony fish conserve energy that can be spent chasing after prey and escaping from predators.
4. a specialized sensory system that extends along each side of a bony fish's body
5. to perceive its position and rate of movement and to detect motionless objects
6. a special gas sac that enables a bony fish to regulate its buoyancy
7. Because the swim bladder of early bony fishes was connected to the throat, the fishes could gulp air to fill the swim bladder.
8. c

SECTION: AMPHIBIANS

1. Caecilians lost their legs during the evolutionary course of adapting to a burrowing existence.
2. Pulmonary veins return oxygen-rich blood from the lungs to the heart. The heart then pumps the blood, at a higher pressure than when it left the lungs, to the body tissues.
3. respiring through the skin, or skin breathing
4. Cutaneous respiration is efficient only when there is a high ratio of skin surface area to body volume.
5. d

Vocabulary Review

ACROSS

2. COUNTERCURRENT
4. OPERCULUM
7. LATERAL
8. LUNG
11. BLADDER

DOWN

1. PULMONARY
3. NEPHRON
5. TELEOST
6. FILAMENTS
9. GILL
10. SEPTUM

Science Skills

INTERPRETING DIAGRAMS

1. *A* is the conus arteriosus, *B* is the atrium, *C* is the sinus venosus, and *D* is the ventricle.
2. to smooth the pulsations in the blood and add more force
3. to reduce the resistance of blood flow into the heart
4. to the gills and then to the rest of the body
5. from the body
6. The walls are thick and muscular, allowing chamber *D* (the ventricle) to force blood to flow through the gills and the rest of the body.
7. *E*, *B*, *D*, *A*, *C*
8. metamorphosis
9. a tadpole
10. Stage *B* (a tadpole) has a tail, lacks legs, breathes with gills, and feeds mostly on algae. An adult frog is tailless, has four legs, breathes with lungs and through its skin, and is carnivorous.
11. Stage *E* is produced when a female frog releases her eggs into the water and a male frog's sperm fertilizes them externally.

Concept Mapping

1. jawless
2. cartilaginous
3. hagfishes
4. sharks or skates
5. skates or sharks
6. ray-finned fishes
7. coelacanth or lungfishes
8. lungfishes or coelacanth

Critical Thinking

1. b	**12.** c
2. a	**13.** e
3. c	**14.** i, f
4. d	**15.** a, j
5. b	**16.** d, b
6. a	**17.** c, h
7. d	**18.** e, g
8. c	**19.** a
9. a	**20.** d
10. d	**21.** b
11. b	**22.** c

Test Prep Pretest

1. c
2. d
3. a
4. c
5. b
6. d
7. b
8. gills
9. scales
10. pulmonary veins
11. vertebral column
12. lungs
13. optic lobe
14. nephrons
15. Urodela
16. countercurrent flow
17. In countercurrent flow, water passing across the gills and blood circulating through the capillaries in the gills flow in opposite directions. This ensures that oxygen diffuses into the blood over the entire length of the capillaries in the gills.
18. Sound causes the tympanic membrane to vibrate. These vibrations are sent to the inner ear through a tiny bone. Ciliated sensory cells in the inner ear detect the vibrations and help the frog maintain its balance.
19. Most lampreys are parasites of other fishes. Hagfishes are scavengers of dead and dying animals on the ocean bottom.
20. The fish uses this system to perceive its position and rate of movement and to detect objects in the water.

Quiz

SECTION: THE FISH BODY

1. b	**6.** c
2. e	**7.** c
3. a	**8.** d
4. f	**9.** a
5. d	**10.** b

SECTION: TODAY'S FISHES

1. d	**6.** c
2. b	**7.** b
3. f	**8.** a
4. e	**9.** c
5. a	**10.** d

SECTION: AMPHIBIANS

1. d	**6.** c
2. b	**7.** d
3. e	**8.** a
4. f	**9.** a
5. a	**10.** c

Chapter Test (General)

1. b	**11.** c
2. d	**12.** d
3. g	**13.** c
4. c	**14.** c
5. f	**15.** a
6. e	**16.** b
7. a	**17.** c
8. d	**18.** b
9. a	**19.** a
10. d	**20.** d

Chapter Test (Advanced)

1. e
2. d
3. a
4. c
5. b
6. f
7. d
8. b
9. d
10. c
11. c
12. a
13. muscles, opercula
14. cutaneous
15. lampreys, hagfishes
16. cloacal

17. scales, age
18. pulmonary
19. metamorphosis
20. nephrons
21. Caecilians do not have legs. They lost their legs during the evolutionary course of adapting to a burrowing existence.
22. Mucous glands embedded in the skin supply a lubricant that keeps the skin moist. This moist surface is necessary for cutaneous respiration.
23. The operculum, or gill cover, permits a bony fish to pump water over the gills, even when the fish is not swimming.
24. The caudal fin thrusts from side to side to propel the fish forward. The dorsal fins prevent the fish from rolling over as it swims. The pectoral fins assist the fish in going up or down, turning sharply, and stopping quickly.
25. Oxygen-poor blood from the body enters the sinus venosus and then the right atrium. Oxygen-rich blood from the lungs enters the left atrium through the pulmonary veins. The blood in both atria enters the ventricle, which pumps the blood to the conus arteriosus and then to the lungs and the body. Oxygen-poor blood tends to go to the lungs. Oxygen-rich blood tends to go to the body.

Lesson Plan

Section: The Fish Body

Pacing

Regular Schedule:	**with lab(s):** N/A	**without lab(s):** 3 days
Block Schedule:	**with lab(s):** N/A	**without lab(s):** 1 1/2 days

Objectives

1. Describe the characteristics common to modern fishes.
2. Summarize how fish obtain oxygen.
3. Summarize how blood circulates through a fish.
4. Contrast how marine and freshwater fishes balance their salt and water content.
5. Describe two methods of reproduction in fishes.

National Science Education Standards Covered

UNIFYING CONCEPTS AND PROCESSES

UCP1: Systems, order, and organization

UCP2: Evidence, models, and explanation

UCP5: Form and function

SCIENCE AS INQUIRY

SI1: Abilities necessary to do scientific inquiry

SI2: Understandings about scientific inquiry

LIFE SCIENCE: BIOLOGICAL EVOLUTION

LSEvol1: Species evolve over time.

LSEvol2: The great diversity of organisms is the result of more than 3.5 billion years of evolution.

LSEvol5: Biological classifications are based on how organisms are related.

LIFE SCIENCE: MATTER, ENERGY, AND ORGANIZATION IN LIVING SYSTEMS

LSMat4: The complexity and organization of organisms accommodates the need for obtaining, transforming, transporting, releasing, and eliminating the matter and energy used to sustain the organism.

LIFE SCIENCE: BEHAVIOR OF ORGANISMS

LSBeh1: Multicellular animals have nervous systems that generate behavior.

LSBeh2: Organisms have behavioral responses to internal changes and to external stimuli.

KEY
SE = Student Edition — **TE** = Teacher Edition
CRF = Chapter Resource File

Block 1

CHAPTER OPENER *(45 minutes)*

_ **Quick Review,** SE. Students answer questions covered in previous sections of the textbook as preparation for the chapter content. **(GENERAL)**

_ **Reading Activity,** SE. After reading each section of the chapter, students review the figures and their captions and write one question that can be answered by each one. **(GENERAL)**

_ **Using the Figure,** TE. Students answer questions about the chapter opener photograph. **(GENERAL)**

_ **Opening Activity**, Jaw Function, TE. Bring several sizes and types of pliers to class. Ask students to decide which tools exert the most force and which are best for picking up small objects. Relate the function of the tools to the function of animals' jaws. **(BASIC)**

Block 2

FOCUS *(5 minutes)*

_ **Bellringer Transparency.** Use this transparency as students enter the classroom and find their seats. **(GENERAL)**

MOTIVATE *(10 minutes)*

_ **Discussion**, TE. Write the following terms on the board: *starfish, sea horse, sea cow, lamprey, eel,* and *stingray.* Ask students to identify which organisms they think are fish. **(GENERAL)**

TEACH *(30 minutes)*

_ **Teaching Transparency, Section Outline.** Use this transparency to give students a framework for the information in this section. **(GENERAL)**

_ **Inclusion Strategies**, TE. Students research and give an oral report on the seahorse.

_ **Integrating Physics and Chemistry**, TE. Ask students where a higher concentration of dissolved oxygen gas is likely to be found, near the surface or deeper in the water.

_ **Teaching Transparency, Respiration in Fishes.** Use this transparency to describe the structure and function of a fish's gills. Point out that blood and water flow in opposite directions. **(GENERAL)**

_ **Teaching Transparency, Fish Heart Structure.** Use this transparency to discuss how blood flows through the heart of a fish. **(GENERAL)**

HOMEWORK

_ **Active Reading Worksheet, The Fish Body, CRF.** Students read a passage related to the section topic and answer questions. **(GENERAL)**

_ **Directed Reading Worksheet, The Fish Body, CRF.** Students complete the exercises in this worksheet to help them understand the material as they read the section. **(BASIC)**

Block 3

TEACH *(35 minutes)*

_ **Demonstration**, TE. Students use an aquarium siphon with a bulb attached to it, a small piece of rubber or glass tubing, and two jars of water to demonstrate the simple tubular heart of ancestral chordates.

_ **Demonstration**, TE. Students use potato slices and one beaker each of regular distilled water and distilled water mixed with salt to show the effects that these solutions would have on a freshwater fish. **(GENERAL)**

_ **Data Lab,** Analyzing Ion Excretion in Fish, SE. Students study a graph to analyze the change in ion concentration of a fish as it travels from one body of water to another. **(GENERAL)**

_ **Datasheets for In-Text Labs, Analyzing Ion Excretion in Fish, CRF.**

CLOSE *(10 minutes)*

_ **Reteaching,** TE. Students review their lists of things they want to know from the Bellringer activity in the TE. **(BASIC)**

_ **Quiz,** TE. Students answer questions that review the section material. **(GENERAL)**

HOMEWORK

_ **Alternative Assessment**, TE. Students make two columns on a sheet of paper and label one column *External characteristics* and the other column *Internal characteristics*. Then they review the chapter and list the major internal and external characteristics of fishes. **(GENERAL)**

_ **Section Review,** SE. Assign questions 1–6 for review, homework, or quiz. **(GENERAL)**

_ **Quiz, CRF.** This quiz consists of ten multiple choice and matching questions that review the section's main concepts. **(BASIC) Also in Spanish.**

Other Resource Options

- **Career**, Aquaculturist, TE. Lead a discussion on aquaculture or fish farming. Operating a fish farm requires knowledge of feeding, infections, and how to maintain a proper environment.
- **Internet Connect.** Students can research Internet sources about Kidneys with SciLinks Code HX4110.
- **Internet Connect.** Students can research Internet sources about Sharks with SciLinks Code HX4163.
- **go.hrw.com.** For worksheets, videos, and other teaching aids related to this chapter, visit the HRW Web site and type in the keyword HX4 FSA.
- **CNN Science in the News, Video Segment 27 Fish Farming.** This video segment is accompanied by a **Critical Thinking Worksheet**.
- **CNN Student News.** Find the latest news, lesson plans, and activities related to important scientific events at **cnnstudentnews.com**.

Lesson Plan

Section: Today's Fishes

Pacing

Regular Schedule:	**with lab(s):** 3 days	**without lab(s):** 2 days
Block Schedule:	**with lab(s):** 1 1/2 days	**without lab(s):** 1 day

Objectives

1. Distinguish between the three general categories of modern fish.
2. Describe the major external and internal characteristics of the yellow perch.
3. Summarize the unique features of bony fishes.

National Science Education Standards Covered

UNIFYING CONCEPTS AND PROCESSES

UCP5: Form and function

SCIENCE AS INQUIRY

SI1: Abilities necessary to do scientific inquiry

SI2: Understandings about scientific inquiry

LIFE SCIENCE: BIOLOGICAL EVOLUTION

LSEvol1: Species evolve over time.

LSEvol2: The great diversity of organisms is the result of more than 3.5 billion years of evolution.

LSEvol5: Biological classifications are based on how organisms are related.

LIFE SCIENCE: MATTER, ENERGY, AND ORGANIZATION IN LIVING SYSTEMS

LSMat4: The complexity and organization of organisms accommodates the need for obtaining, transforming, transporting, releasing, and eliminating the matter and energy used to sustain the organism.

LIFE SCIENCE: BEHAVIOR OF ORGANISMS

LSBeh1: Multicellular animals have nervous systems that generate behavior.

LSBeh2: Organisms have behavioral responses to internal changes and to external stimuli.

KEY
SE = Student Edition
TE = Teacher Edition
CRF = Chapter Resource File

Block 4

FOCUS *(5 minutes)*

_ **Bellringer Transparency.** Use this transparency as students enter the classroom and find their seats. **(GENERAL)**

MOTIVATE *(10 minutes)*

_ **Discussion**, TE. Show students pictures of various fish, and explain that they are classified into three distinct groups. Have students try to determine which fish belong in the same groups and why. Discuss how the structural characteristics are often important in classifying organisms and how the fishes are good examples of this. **(BASIC)**

TEACH *(30 minutes)*

_ **Teaching Transparency, Section Outline.** Use this transparency to give students a framework for the information in this section. **(GENERAL)**

_ **Teaching Tip**, Natural Selection of Body Shape, TE. To look for similarities in body shape, have students diagram the profile of several living and nonliving things that move quickly through the water. Ask students why these shapes occur in things that need to move rapidly through the water. **(GENERAL)**

_ **Integrating Physics and Chemistry**, TE. Ask students to identify the medium that perpetuates the waves from an object to the lateral line of a fish. **(GENERAL)**

_ **Teaching Transparency, Lateral Line in Bony Fish.** Use this transparency to describe the structure and function of the lateral line. Point out that the lateral line contains sensory cells that help a fish perceive its position in the water. **(GENERAL)**

HOMEWORK

_ **Directed Reading Worksheet, Today's Fishes, CRF.** Students complete the exercises in this worksheet to help them understand the material as they read the section. **(BASIC)**

_ **Active Reading Worksheet, Today's Fishes, CRF.** Students read a passage related to the section topic and answer questions. **(GENERAL)**

_ **Activity**, Fish and Diet, TE. Have students research commercial fish to determine where they were caught or raised; their habitats; their nutrient, vitamin, and mineral contents; and any other interesting characteristics of the fish. **(GENERAL)**

Block 5

TEACH *(35 minutes)*

_ **Teaching Transparency, External Structures of Fish.** Use this transparency to give students a framework for the information in this section. **(GENERAL)**

_ **Teaching Transparency, Internal Structures of Fish.** Use this transparency to give students a framework for the information in this section. **(GENERAL)**

_ **Up Close,** Yellow Perch, SE. Use the teaching strategies and discussion questions in the TE to help students better understand the external and internal structures of the yellow perch. **(GENERAL)**

_ **Quick Lab,** Modeling the Action of a Swim Bladder, SE. Students model a swim bladder using raisins and a carbonated beverage. **(GENERAL)**

_ **Datasheets for In-Text Labs, Modeling the Action of a Swim Bladder, CRF.**

CLOSE *(10 minutes)*

_ **Reteaching,** TE. Students work in small groups to make a table that compares each type of fish they have studied in this section. Have each student copy his or her group's table and place it is his or her portfolio. **(BASIC)**

_ **Quiz,** TE. Students answer questions that review the section material. **(GENERAL)**

HOMEWORK

_ **Alternative Assessment,** TE. Students choose a favorite fish and then explain to the class how the fish's characteristics enable it to survive and give it an advantage over other types of fishes. **(GENERAL)**

_ **Section Review,** SE. Assign questions 1–4 for review, homework, or quiz. **(GENERAL)**

_ **Quiz, CRF.** This quiz consists of ten multiple choice and matching questions that review the section's main concepts. **(BASIC) Also in Spanish.**

Optional Block

LAB *(45 minutes)*

_ **Exploration Lab, Observing Structure and Behavior in Fishes, CRF.** Students observe the structural and behavioral adaptations of several species of fish at an aquarium. They then consider how these adaptations help each species survive in its habitat. **(GENERAL)**

Other Resource Options

_ **Integrating Physics and Chemistry**, TE. Ask what happens to a fish's buoyancy when the swim bladder increases or decreases in size. Also ask students whether a fish's swim bladder increases or decreases in size if a fish moves into deeper water. **(GENERAL)**

- **Internet Connect.** Students can research Internet sources about Kidneys with SciLinks Code HX4110.
- **Internet Connect.** Students can research Internet sources about Sharks with SciLinks Code HX4163.
- **go.hrw.com.** For worksheets, videos, and other teaching aids related to this chapter, visit the HRW Web site and type in the keyword HX4 FSA.
- **CNN Student News.** Find the latest news, lesson plans, and activities related to important scientific events at **cnnstudentnews.com**.

Lesson Plan

Section: Amphibians

Pacing

Regular Schedule: **with lab(s):** 5 days **without lab(s):** 2 days

Block Schedule: **with lab(s):** 2 1/2 days **without lab(s):** 1 day

Objectives

1. Summarize the characteristics common to all modern amphibians.
2. Compare the three orders of living amphibians.
3. Describe the major external and internal characteristics of the leopard frog.

National Science Education Standards Covered

UNIFYING CONCEPTS AND PROCESSES

UCP1: Systems, order, and organization

UCP2: Evidence, models, and explanation

UCP5: Form and function

SCIENCE AS INQUIRY

SI1: Abilities necessary to do scientific inquiry

SI2: Understandings about scientific inquiry

LIFE SCIENCE: BIOLOGICAL EVOLUTION

LSEvol1: Species evolve over time.

LSEvol2: The great diversity of organisms is the result of more than 3.5 billion years of evolution.

LSEvol5: Biological classifications are based on how organisms are related.

LIFE SCIENCE: MATTER, ENERGY, AND ORGANIZATION IN LIVING SYSTEMS

LSMat4: The complexity and organization of organisms accommodates the need for obtaining, transforming, transporting, releasing, and eliminating the matter and energy used to sustain the organism.

LIFE SCIENCE: BEHAVIOR OF ORGANISMS

LSBeh1: Multicellular animals have nervous systems that generate behavior.

LSBeh2: Organisms have behavioral responses to internal changes and to external stimuli

KEY

SE = Student Edition **TE** = Teacher Edition

CRF = Chapter Resource File

Block 6

FOCUS *(5 minutes)*

_ **Bellringer Transparency.** Use this transparency as students enter the classroom and find their seats. **(GENERAL)**

MOTIVATE *(10 minutes)*

_ **Activity**, Mitosis, TE. Students classify different amphibians among three orders. **(GENERAL)**

TEACH *(30 minutes)*

_ **Teaching Transparency, Section Outline.** Use this transparency to give students a framework for the information in this section. **(GENERAL)**

_ **Skill Builder**, Reading Organizer, TE. Students draw a Venn diagram that compares the characteristics of fishes and amphibians. A sample Venn diagram is provided in the TE. **(GENERAL)**

_ **Teaching Transparency, Fish and Amphibian Circulation.** Use this transparency to compare the single-loop circulation found in most fishes with the double-loop circulation found in amphibians and lungfishes. **(GENERAL)**

_ **Teaching Transparency, Amphibian Heart Structure.** Use this transparency to trace the flow of blood through the amphibian heart. **(GENERAL)**

_ **Teaching Tip**, Comparing Heart Anatomy, TE. Contrast the heart of a fish with that of an amphibian. **(GENERAL)**

HOMEWORK

_ **Directed Reading Worksheet, Amphibians, CRF.** Students complete the exercises in this worksheet to help them understand the material as they read the section. **(BASIC)**

_ **Active Reading Worksheet, Amphibians, CRF.** Students read a passage related to the section topic and answer questions. **(GENERAL)**

Block 7

TEACH *(35 minutes)*

_ **Inclusion Strategies**, TE. Students create a poster showing the differences between toads and frogs.

_ **Teaching Transparency, Life Cycle of a Frog.** Use this transparency to describe the frog's life cycle. Tell students that the eggs of frogs and toads are unprotected and jellylike. If the eggs are removed from water, they will dry out, and the embryos inside will die. **(GENERAL)**

_ **Up Close**, Leopard Frog, SE. Use the teaching strategies and discussion questions in the TE to help students better understand the external and internal structures of the yellow perch. **(GENERAL)**

_ **Teaching Transparency, External Structure of a Frog.** Use this transparency to discuss the tympanic membrane, eye, and skin of the leopard frog . **(GENERAL)**

_ **Teaching Transparency, Internal Structure of a Frog.** Use this transparency to discuss the brain, tongue and jaw, cloaca, and reproductive organs of the leopard frog. **(GENERAL).**

CLOSE *(10 minutes)*

_ **Reteaching,** TE. Students make a list of characteristics that distinguish amphibians from fish. **(BASIC)**

_ **Quiz,** TE. Students answer questions that review the section material. **(GENERAL)**

HOMEWORK

_ **Alternative Assessment,** TE. Students work in groups to prepare a poster showing the frog's life cycle. Have them include drawings of each of the four stages of a frog's life and a paragraph that explains what is illustrated by the drawings and what they have learned. **(GENERAL)**

_ **Section Review,** SE. Assign questions 1–7 for review, homework, or quiz. **(GENERAL)**

_ **Science Skills Worksheet, CRF.** Students interpret diagrams of a fish heart and the frog life cycle. **(GENERAL)**

_ **Quiz, CRF.** This quiz consists of ten multiple choice and matching questions that review the section's main concepts. **(BASIC) Also in Spanish.**

_ **Modified Worksheet, One-Stop Planner**. This worksheet has been specially modified to reach struggling students. **(BASIC)**

_ **Critical Thinking Worksheet, CRF.** Students answer analogy-based questions that review the section's main concepts and vocabulary. **(ADVANCED)**

Optional Blocks

LABS *(135 minutes)*

_ **Exploration Lab,** Observing a Live Frog, SE. Students examine a live frog in both a terrestrial environment and in an aquatic environment. **(GENERAL)**

_ **Datasheets for In-Text Labs, Observing a Live Frog, CRF.**

_ **Skills Practice Lab, Frog Dissection, CRF.** Students dissect a frog in order to observe the external and internal structures of frog anatomy. **(GENERAL)**

Other Resource Options

_ **Teaching Tip**, Tadpole Metamorphosis, TE. Have students observe tadpole development in the classroom from the hind-limb budding stage. Inform them of internal changes as well as external changes. **(BASIC)**

_ **Supplemental Reading, Through a Window, One-Stop Planner.** Students read the book and answer questions. **(ADVANCED)**

_ **Internet Connect.** Students can research Internet sources about Circulation in Amphibians with SciLinks Code HX4043.

_ **go.hrw.com.** For worksheets, videos, and other teaching aids related to this chapter, visit the HRW Web site and type in the keyword HX4 FSA.

_ **CNN Student News.** Find the latest news, lesson plans, and activities related to important scientific events at **cnnstudentnews.com**.

Lesson Plan

End-of-Chapter Review and Assessment

Pacing

Regular Schedule: 2 days

Block Schedule: 1 day

KEY	
SE = Student Edition	TE = Teacher Edition
CRF = Chapter Resource File	

Block 8

REVIEW *(45 minutes)*

_ **Study Zone,** SE. Use the Study Zone to review the Key Concepts and Key Terms of the chapter and prepare students for the Performance Zone questions. **(GENERAL)**

_ **Performance Zone,** SE. Assign questions to review the material for this chapter. Use the assignment guide to customize review for sections covered. **(GENERAL)**

_ **Teaching Transparency, Concept Mapping.** Use this transparency to review the concept map for this chapter. **(GENERAL)**

Block 9

ASSESSMENT *(45 minutes)*

_ **Chapter Test, Fishes and Amphibians, CRF.** This test contains 20 multiple choice and matching questions keyed to the chapter's objectives. **(GENERAL) Also in Spanish.**

_ **Chapter Test, Fishes and Amphibians, CRF.** This test contains 25 questions of various formats, each keyed to the chapter's objectives. **(ADVANCED)**

_ **Modified Chapter Test, One-Stop Planner.** This test has been specially modified to reach struggling students. **(BASIC)**

Other Resource Options

_ **Vocabulary Review Worksheet, CRF.** Use this worksheet to review the chapter vocabulary. **(GENERAL) Also in Spanish.**

_ **Test Prep Pretest, CRF.** Use this pretest to review the main content of the chapter. Each question is keyed to a section objective. **(GENERAL) Also in Spanish.**

_ **Test Item Listing for ExamView® Test Generator, CRF.** Use the Test Item Listing to identify questions to use in a customized homework, quiz, or test.

_ **ExamView® Test Generator, One-Stop Planner.** Create a customized homework, quiz, or test using the HRW Test Generator program.

TEST ITEM LISTING

Fishes and Amphibians

TRUE/FALSE

1. ____ Members of the class Osteichthyes have skeletons of cartilage.
Answer: False Difficulty: I Section: 1 Objective: 1

2. ____ In modern fishes, gas exchange takes place through membranes of their gills.
Answer: True Difficulty: I Section: 1 Objective: 2

3. ____ All fishes have an internal skeleton.
Answer: True Difficulty: I Section: 1 Objective: 1

4. ____ All fishes have a bony skeleton.
Answer: False Difficulty: I Section: 1 Objective: 1

5. ____ Almost all fishes have single-loop circulation.
Answer: True Difficulty: I Section: 1 Objective: 1

6. ____ In fishes, oxygenated blood passes from the gills to the heart and then to the rest of the body.
Answer: False Difficulty: I Section: 1 Objective: 2

7. ____ During countercurrent flow, water flows over the gills of fish in the same direction that blood flows within their gills.
Answer: False Difficulty: I Section: 1 Objective: 2

8. ____ In a fish, countercurrent flow prevents a steady supply of oxygen from reaching the gill filaments.
Answer: False Difficulty: I Section: 1 Objective: 2

9. ____ The gills of bony fishes are extremely efficient respiratory organs.
Answer: True Difficulty: I Section: 1 Objective: 2

10. ____ The circulatory system of fishes includes a simple tubular heart.
Answer: False Difficulty: I Section: 1 Objective: 3

11. ____ Fish need a stronger pump than a simple tubular heart provides due to the capillaries in the gills, which create resistance to the flow of blood.
Answer: True Difficulty: I Section: 1 Objective: 3

12. ____ Contractions of a fish heart's ventricle pump blood toward the gills.
Answer: True Difficulty: I Section: 1 Objective: 3

13. ____ The sinus venosus acts like a collection chamber in the fish heart.
Answer: True Difficulty: I Section: 1 Objective: 3

14. ____ Since they live in salt water, marine fishes do not have a problem maintaining the proper balance of water and salt in their bodies.
Answer: False Difficulty: I Section: 1 Objective: 4

15. ____ Minimizing water loss is a key evolutionary challenge in some fishes.
Answer: True Difficulty: I Section: 1 Objective: 4

16. ____ Nephridia are complex organs in fish that regulate the body's salt and water balance.
Answer: False Difficulty: I Section: 1 Objective: 4

17. ____ In fish, excess water and bodily wastes leave the kidneys in the form of uric acid.
Answer: False Difficulty: I Section: 1 Objective: 4

18. ____ Marine fishes excrete large amounts of dilute urine.
Answer: False Difficulty: I Section: 1 Objective: 4

19. ____ Marine fishes excrete small amounts of urine.
Answer: True Difficulty: I Section: 1 Objective: 4

20. ____ Fishes carry out both asexual and sexual reproduction.
Answer: False Difficulty: I Section: 1 Objective: 5

21. ____ Most fishes have external fertilization.
Answer: True Difficulty: I Section: 1 Objective: 5

22. ____ Some fishes give birth to live young.
Answer: True Difficulty: I Section: 1 Objective: 5

23. ____ Bony fishes have a swim bladder.
Answer: True Difficulty: I Section: 2 Objective: 3

24. ____ To fill their swim bladders, bony fishes have to come to the surface to gulp air.
Answer: False Difficulty: I Section: 2 Objective: 3

25. ____ All fishes have cartilaginous skeletons.
Answer: False Difficulty: I Section: 2 Objective: 1

26. ____ Lampreys and hagfishes are jawless fishes.
Answer: True Difficulty: I Section: 2 Objective: 1

27. ____ Lampreys and hagfishes have scaled bodies and multiple paired fins.
Answer: False Difficulty: I Section: 2 Objective: 1

28. ____ Lampreys are also known as the "vultures of the sea."
Answer: False Difficulty: I Section: 2 Objective: 1

29. ____ The skeletons of sharks and rays are made of cartilage.
Answer: True Difficulty: I Section: 2 Objective: 1

30. ____ The scales and teeth of sharks are very similar in structure.
Answer: True Difficulty: I Section: 2 Objective: 1

31. ____ Bony fishes gulp air to fill their swim bladders.
Answer: False Difficulty: I Section: 2 Objective: 3

32. ____ The fins of yellow perch are used to draw water into the mouth and over the gills.
Answer: False Difficulty: I Section: 2 Objective: 2

33. ____ Movements of the opercula draw water into the yellow perch's mouth.
Answer: True Difficulty: I Section: 2 Objective: 2

34. ____ The structure of the digestive system of yellow perch is very similar to that of all vertebrates.
Answer: True Difficulty: I Section: 2 Objective: 2

35. ____ Bony fishes must remain swimming if they are to respire.
Answer: False Difficulty: I Section: 2 Objective: 3

36. ____ A bony fish must swim forward with its mouth open to move water over its gills.
Answer: False Difficulty: I Section: 2 Objective: 3

37. ____ The lateral line system of bony fishes is a specialized sensory system.
Answer: True Difficulty: I Section: 2 Objective: 3

38. ____ Because their skin is watertight, amphibians can live anywhere on dry land.
Answer: False Difficulty: I Section: 3 Objective: 1

39. ____ In amphibians, the blood that is pumped from the heart to the body is completely oxygenated.
Answer: False Difficulty: I Section: 3 Objective: 1

40. ____ Frog eggs must be laid in a wet or moist environment.
Answer: True Difficulty: I Section: 3 Objective: 2

41. ____ In most amphibians, fertilization takes place externally.
Answer: True Difficulty: I Section: 3 Objective: 1

42. ____ The skin of most amphibians functions as a respiratory organ.
Answer: True Difficulty: I Section: 3 Objective: 1

43. ____ Pulmonary veins carry oxygenated blood to the heart of an amphibians.
Answer: True Difficulty: I Section: 3 Objective: 1

44. ____ A frog must remain in water in order to survive.
Answer: False Difficulty: I Section: 3 Objective: 2

45. ____ Caecilians are tropical amphibians with four legs and a tail.
Answer: False Difficulty: I Section: 3 Objective: 2

46. ____ Frogs and toads are in the only order of herbivorous amphibians.
Answer: False Difficulty: I Section: 3 Objective: 2

47. ____ Anurans are found in desert as well as mountain environments.
Answer: True Difficulty: I Section: 3 Objective: 2

48. ____ Unlike frog and toad larvae, salamanders do not undergo a dramatic metamorphosis.
Answer: True Difficulty: I Section: 3 Objective: 2

49. ____ Caecilians are found in every environment on Earth.
Answer: False Difficulty: I Section: 3 Objective: 2

50. ____ Leopard frogs hear well only when they are under water.
Answer: False Difficulty: I Section: 3 Objective: 3

51. ____ The eyelids of frogs protect the eyes from dust.
Answer: True Difficulty: I Section: 3 Objective: 3

MULTIPLE CHOICE

52. All fishes have a(n)
a. internal skeleton.
b. external skeleton.
c. cartilaginous skeleton.
d. bony skeleton.
Answer: A Difficulty: I Section: 1 Objective: 1

53. Sea horses are
a. caecilians.
b. fishes.
c. amphibians.
d. salamanders.
Answer: B Difficulty: I Section: 1 Objective: 1

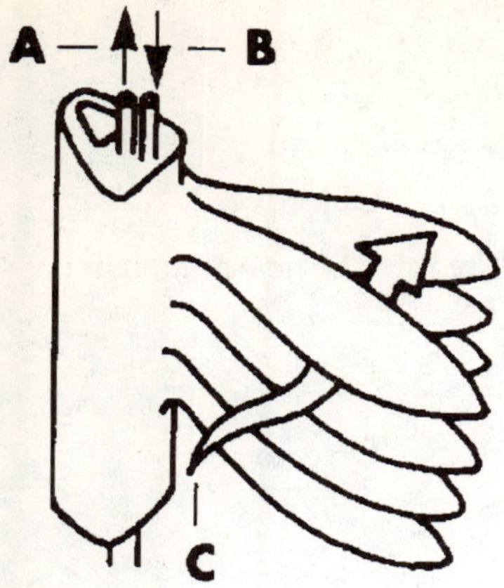

54. Refer to the illustration above. The structure shown is a
    a. lung.
    b. lateral line.
    c. gill.
    d. ventricle.

    Answer: C Difficulty: II Section: 1 Objective: 2

55. Refer to the illustration above. Which arrow indicates the direction of water flow?
    a. arrow A
    b. arrow B
    c. arrow C
    d. None of the arrows show the direction of water flow.

    Answer: C Difficulty: II Section: 1 Objective: 2

56. Refer to the illustration above. In which of the following organisms might you expect to find the structure illustrated?
    a. frog
    b. salamander
    c. newt
    d. goldfish

    Answer: D Difficulty: II Section: 1 Objective: 2

57. The concurrent flow of water and blood found in the gills of fishes
    a. allows blood and water to flow in the same direction.
    b. ensures that oxygen diffuses into the blood over the whole length of the blood vessels in the gills.
    c. results in an uneven supply of oxygen reaching the blood vessels in the gills.
    d. inhibits the diffusion of oxygen and carbon dioxide between the blood and the water.

    Answer: B Difficulty: I Section: 1 Objective: 2

58. When the blood enters the gill filaments in a fish,
    a. its carbon dioxide content is low and oxygen content is high.
    b. its carbon dioxide content is low and oxygen content is low.
    c. its carbon dioxide content is high and oxygen content is high.
    d. its carbon dioxide content is high and oxygen content is low.

    Answer: D Difficulty: I Section: 1 Objective: 2

59. Generally, fishes obtain oxygen through their
    a. skin.
    b. lungs.
    c. gills.
    d. mantles.

    Answer: C Difficulty: I Section: 1 Objective: 2

60. folds : lungs ::
   a. fire : wood
   b. fish : gill filaments
   c. oxygen : atmosphere
   d. gill filaments : gills

   Answer: D Difficulty: II Section: 1 Objective: 2

61. A collection chamber that reduces the resistance of blood flow into the heart of a fish is called the
   a. sinus venosus.
   b. ventricle.
   c. conus arteriosus.
   d. atrium.

   Answer: A Difficulty: I Section: 1 Objective: 3

62. In a fish, oxygenated blood moves from the gills to the
   a. heart.
   b. sinus venosus.
   c. body.
   d. conus arteriosus.

   Answer: C Difficulty: I Section: 1 Objective: 3

63. In a fish, the chamber that pumps blood to the gills is the
   a. sinus venosus.
   b. septum.
   c. atrium.
   d. ventricle.

   Answer: D Difficulty: I Section: 1 Objective: 3

64. The amount of urine a fish produces depends primarily on the
   a. time of year.
   b. size of the fish.
   c. environment in which the fish lives.
   d. diet of the fish.

   Answer: C Difficulty: I Section: 1 Objective: 4

65. To make up for water they lose from their body, marine fishes
   a. drink sea water.
   b. actively pump excess ions into their bodies.
   c. take in water through osmosis.
   d. actively take in salts from their environment.

   Answer: A Difficulty: I Section: 1 Objective: 4

66. freshwater fish's kidney : large amount of urine ::
   a. nephron : kidney
   b. marine fish's kidney : small amount of urine
   c. milk : kitten
   d. salt water intake : absorption

   Answer: B Difficulty: II Section: 1 Objective: 4

67. The eggs of sharks are
   a. covered by a watertight outer covering.
   b. fertilized internally.
   c. released from the body before fertilization.
   d. fertilized externally.

   Answer: B Difficulty: I Section: 1 Objective: 5

68. The release of eggs and sperm near each other into the water occurs during
   a. growth.
   b. development.
   c. accretion.
   d. spawning.

   Answer: D Difficulty: I Section: 1 Objective: 5

69. Lampreys and hagfishes have
    a. jaws.
    b. paired fins.
    c. a rigid skeleton.
    d. a notochord in all stages of their life.

    Answer: D   Difficulty: I   Section: 2   Objective: 1

70. Jawless fishes include
    a. lampreys and sharks.
    b. sharks and rays.
    c. hagfishes and coelacanths.
    d. lampreys and hagfishes.

    Answer: D   Difficulty: I   Section: 2   Objective: 1

71. Most lampreys are
    a. autotrophs.
    b. herbivores.
    c. parasites.
    d. amphibians.

    Answer: C   Difficulty: I   Section: 2   Objective: 1

72. Skates and rays are
    a. lobe-finned fishes.
    b. jawless fishes.
    c. bony fishes.
    d. cartilaginous fishes.

    Answer: D   Difficulty: I   Section: 2   Objective: 1

73. Sharks have all of the following *except*
    a. placoid scales.
    b. internal fertilization.
    c. a swim bladder.
    d. a cartilaginous skeleton.

    Answer: C   Difficulty: I   Section: 2   Objective: 1

74. jaws : lampreys ::
    a. cartilage : shark
    b. fins : jawed fishes
    c. teeth : jaws
    d. paired fins : jawless fishes

    Answer: D   Difficulty: II   Section: 2   Objective: 1

Three Types of Fishes

75. Refer to the illustration above. Fish A
    a. has skin covered by overlapping structures called scales.
    b. has many small scales embedded in its skin.
    c. feeds parasitically on other fish.
    d. has a skeleton made of bone.

    Answer: B   Difficulty: II   Section: 2   Objective: 1

76. Refer to the illustration above. Fish B
    a. has skin covered by overlapping structures called scales.
    b. has many small scales embedded in its skin.
    c. feeds parasitically on other fish.
    d. has a skeleton made of cartilage.

    Answer: A   Difficulty: II   Section: 2   Objective: 2

77. Refer to the illustration above. Fish C
    a. has skin covered by overlapping structures called scales.
    b. has many small scales embedded in its skin.
    c. feeds parasitically on other fish.
    d. has a skeleton made of bone.

    Answer: C Difficulty: II Section: 2 Objective: 1

78. In the yellow perch,
    a. internal fertilization takes place.
    b. external fertilization takes place.
    c. internal development takes place.
    d. All of the above

    Answer: B Difficulty: I Section: 2 Objective: 2

79. A major difference between bony fishes and sharks is
    a. the presence of a swim bladder in most bony fishes.
    b. the presence of opercula in sharks.
    c. the absence of jaws in bony fishes.
    d. All of the above

    Answer: A Difficulty: I Section: 2 Objective: 1

80. Which of the following is *not* a category of fish?
    a. Osteichthyes
    b. Chondrichthyes
    c. Urodela
    d. Agnatha

    Answer: C Difficulty: I Section: 2 Objective: 1

81. The two major groups of bony fishes are the lobe-finned fishes and the
    a. lungfishes.
    b. ray-finned fishes.
    c. coelacanths.
    d. lampreys.

    Answer: B Difficulty: I Section: 2 Objective: 3

82. Members of the category Osteichthyes
    a. have skeletons made of bone.
    b. do not have jaws.
    c. include the rays and skates.
    d. All of the above

    Answer: A Difficulty: I Section: 2 Objective: 3

83. The coelacanth is a
    a. lungfish.
    b. cartilaginous fish.
    c. ray-finned fish.
    d. lobe-finned fish.

    Answer: D Difficulty: I Section: 2 Objective: 3

84. The swim bladder
    a. stores air for breathing.
    b. contains antibodies.
    c. is found in all amphibians.
    d. allows fishes to become more buoyant.

    Answer: D Difficulty: I Section: 2 Objective: 3

85. The opercula of a bony fish
    a. are part of the skeletal system.
    b. are adaptations for rapid swimming.
    c. cover the gills in fish.
    d. None of the above

    Answer: C Difficulty: I Section: 2 Objective: 3

86. The gills of bony fishes
    a. give them buoyancy.
    b. function well in or out of water.
    c. are composed of scales.
    d. are housed in chambers on each side of the head.

    Answer: D Difficulty: I Section: 2 Objective: 3

External Structure of a Yellow Perch

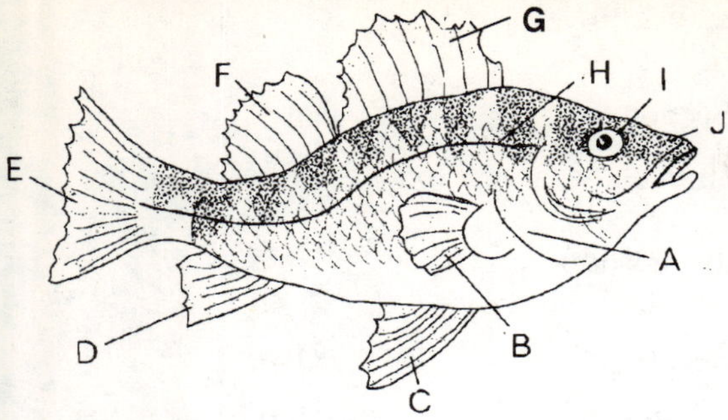

87. Refer to the illustration above. To move forward, the fish uses the fin labeled
   a. B.
   b. C.
   c. E.
   d. F.

   Answer: C Difficulty: II Section: 2 Objective: 2

88. Refer to the illustration above. The structure labeled A, which draws water into the mouth of the fish, is the
   a. lateral line.
   b. cerebrum.
   c. oviduct.
   d. operculum.

   Answer: D Difficulty: II Section: 2 Objective: 2

Structure of the Lateral Line of a Fish

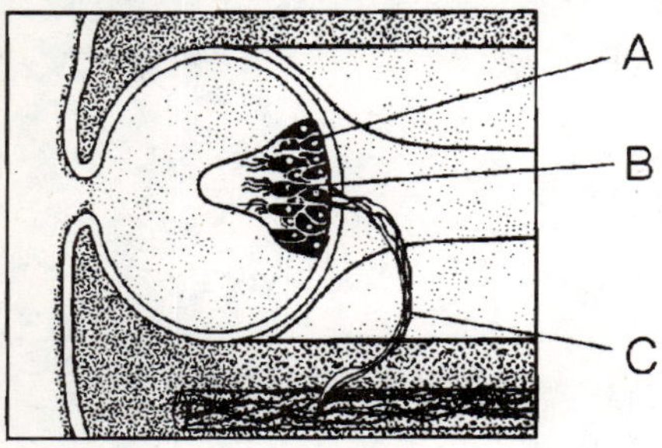

89. Refer to the illustration above. The structure labeled C is a
   a. canal.
   b. sinus venosus.
   c. nephron.
   d. nerve.

   Answer: D Difficulty: II Section: 2 Objective: 3

90. lateral line : motion detector ::
   a. swim bladder : roof on a house
   b. operculum : wheel
   c. operculum : water pump
   d. fin : microwave oven

   Answer: C Difficulty: II Section: 2 Objective: 3

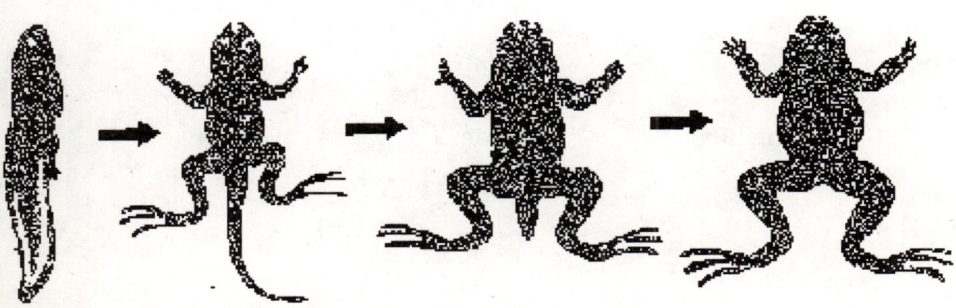

91. Refer to the diagram above. The process illustrated is
    a. amniocentesis.
    b. metamorphosis.
    c. evolution.
    d. synapsis.

    Answer: B Difficulty: II Section: 3 Objective: 2

92. Although adapted to land, most amphibians must have access to a watery environment to
    a. obtain food.
    b. excrete wastes.
    c. reproduce.
    d. All of the above

    Answer: C Difficulty: I Section: 3 Objective: 1

93. In amphibians, gases are exchanged through lung breathing and through the
    a. heart.
    b. air bladder.
    c. lateral line system.
    d. skin.

    Answer: D Difficulty: I Section: 3 Objective: 1

94. Frogs, like most amphibians,
    a. must live in moist areas.
    b. return to the water to reproduce.
    c. have long tails as adults.
    d. belong to the order Urodela.

    Answer: B Difficulty: I Section: 3 Objective: 2

95. During metamorphosis in frogs,
    a. lungs replace gills.
    b. limbs develop.
    c. the tail disappears.
    d. All of the above

    Answer: D Difficulty: I Section: 3 Objective: 2

96. The amphibian heart
    a. pumps only deoxygenated blood.
    b. has four chambers.
    c. pumps only oxygenated blood.
    d. pumps both deoxygenated and oxygenated blood.

    Answer: D Difficulty: I Section: 3 Objective: 1

97. Amphibians without tails are classified in the order
    a. Agnatha.
    b. Anura.
    c. Urodela.
    d. Osteichthyes.

    Answer: B Difficulty: I Section: 3 Objective: 2

98. Newts and salamanders are amphibians in the order
    a. Apoda.
    b. Anura.
    c. Urodela.
    d. Agnatha.

    Answer: C Difficulty: I Section: 3 Objective: 2

99. Amphibians that have slender bodies and no limbs are classified as
    a. anurans.
    b. caecilians.
    c. salamanders.
    d. newts.

    Answer: B Difficulty: I Section: 3 Objective: 2

100. Adult frogs and toads are
    a. herbivores.
    b. omnivores.
    c. parasites.
    d. carnivores.

    Answer: D Difficulty: I Section: 3 Objective: 2

101. tympanic membrane : hearing ::
  a. cloaca : breathing
  b. frog teeth : chewing food
  c. frog tongue : vocalization
  d. frog tongue : catching prey

  Answer: D Difficulty: II Section: 3 Objective: 3

102. Because of their tympanic membrane, leopard frogs can hear well in
  a. both water and air.
  b. water but not in air.
  c. air but not in water.
  d. None of the above

  Answer: A Difficulty: I Section: 3 Objective: 3

COMPLETION

103. During ___________________ -loop circulation, blood is pumped from the heart to the capillaries in the gills, and then to the rest of the body before the blood returns to the heart.

  Answer: single Difficulty: II Section: 1 Objective: 3

104. The vertebral column surrounds the ___________________ ___________________ in fishes.

  Answer: spinal cord Difficulty: I Section: 1 Objective: 1

105. The process of ___________________ is responsible for the exchange of gases across the gills of a fish.

  Answer: diffusion Difficulty: I Section: 1 Objective: 2

106. Each gill in a fish is made up of rows of gill ___________________ that hang like curtains.

  Answer: filaments Difficulty: II Section: 1 Objective: 2

107. The tiny ___________________ in a fish's gills create resistance to the flow of blood.

  Answer: capillaries Difficulty: I Section: 1 Objective: 3

108. In fishes, the simple tubular pump of early chordates is replaced with a(n) ___________________ -pump heart.

  Answer: chamber Difficulty: II Section: 1 Objective: 3

109. Marine bony fishes lose water continuously by ___________________ to the saltier water in which they swim.

  Answer: osmosis Difficulty: I Section: 1 Objective: 4

110. A kidney is made up of individual units called ___________________ that filter the blood.

  Answer: nephrons Difficulty: II Section: 1 Objective: 4

111. For most fishes, fertilization takes place ___________________.

  Answer: externally Difficulty: II Section: 1 Objective: 5

112. Lampreys are ___________________ on other living fishes.

  Answer: parasitic Difficulty: II Section: 2 Objective: 1

113. The mineral ___________________ ___________________ strengthens the cartilaginous skeletons of sharks and rays.

  Answer: calcium carbonate Difficulty: I Section: 2 Objective: 1

114. Yellow perch produce ___________________ during their breeding season in the spring.

  Answer: gametes Difficulty: II Section: 2 Objective: 2

115. Within the brain of the yellow perch, the __________ __________ receives information from the eyes.
Answer: optic lobe Difficulty: II Section: 2 Objective: 2

116. The __________ are the most advanced of the ray-finned bony fishes.
Answer: teleosts Difficulty: II Section: 2 Objective: 3

117. The __________ __________ gives most bony fishes buoyancy.
Answer: swim bladder Difficulty: I Section: 2 Objective: 3

118. Amphibians pump oxygen-rich blood to their tissues at a high pressure because of their __________ -loop circulatory system.
Answer: double Difficulty: II Section: 3 Objective: 1

119. The changes that transform a tadpole into an adult frog are called __________.
Answer: metamorphosis Difficulty: I Section: 3 Objective: 2

120. In some species of amphibians, a(n) __________ partly or completely divides the atrium into right and left halves.
Answer: septum Difficulty: II Section: 3 Objective: 1

121. Frogs and toads are amphibians in the class __________.
Answer: Anura Difficulty: II Section: 3 Objective: 2

122. Salamanders and __________ are amphibians with elongated bodies and long tails.
Answer: newts Difficulty: II Section: 3 Objective: 2

123. In leopard frogs, the eardrum is called the __________ __________.
Answer: tympanic membrane Difficulty: II Section: 3 Objective: 3

ESSAY

124. Why is maintaining water balance a different process for a freshwater fish than for a marine fish?
Answer:
Sea water has a much higher salt concentration than the cells of marine fishes, so marine fishes continually lose water by osmosis. Marine fishes must drink sea water and excrete only small amounts of urine. Freshwater fishes continually take in water by osmosis and excrete large amounts of dilute urine.
Difficulty: III Section: 1 Objective: 4

125. How is the body design of bony fishes and sharks better adapted for swimming than the body design of primitive fishes?
Answer:
Most bony fishes and sharks have streamlined bodies that are well adapted for movement through the water. Both have a strong, light skeleton, and bony fishes have an assortment of movable fins that aid their swimming.
Difficulty: III Section: 2 Objective: 1

126. Explain the importance of the swim bladder to bony fishes.
Answer:
The swim bladder is a gas-filled sac that gives the fishes control of their buoyancy. By regulating the amount of gas in the swim bladder, bony fishes can change or maintain their depth in the water without expending energy through swimming.
Difficulty: III Section: 2 Objective: 3

127. Why are frogs often found near water?

Answer:

Frogs pass through a larval stage in water before moving to land, and adult frogs return to water to lay eggs or release sperm.

Difficulty: II Section: 3 Objective: 2

9997266110 1 2 3 4 5 6